Show What You Know® on the COMMON CORE

3

Assessing Student Knowledge of the Common Core State Standards (CCSS)

Reading
Mathematics

Name: _____

Published by:

Show What You Know® Publishing
www.showwhatyouknowpublishing.com

Distributed by:
Lorenz Educational Press, a Lorenz company
P.O. Box 802
Dayton, OH 45401-0802
www.LorenzEducationalPress.com

Copyright © 2011, 2012 by Show What You Know® Publishing
All rights reserved.

No part of this book, including interior design, cover design, and icons, may be reproduced or transmitted in any form, by any means (electronic, photocopying, recording, or otherwise).

Standards are from the Common Core State Standards Initiative Web site at www.corestandards.org dated 2011.

Printed in the United States of America

ISBN: 978-1-5923-0452-3

Limit of Liability/Disclaimer of Warranty: The authors and publishers have used their best efforts in preparing this book. Show What You Know® Publishing and the authors make no representations or warranties with respect to the contents of this book and specifically disclaim any implied warranties and shall in no event be liable for any loss of any kind including but not limited to special, incidental, consequential, or other damages.

Acknowledgements

Show What You Know® Publishing acknowledges the following for their efforts in making this assessment material available for students, parents, and teachers:

Cindi Englefield, President/Publisher
Eloise Boehm-Sasala, Vice President/Managing Editor
Jennifer Harney, Editor/Illustrator

About the Contributors

The content of this book was written BY teachers FOR teachers and students and was designed specifically for the Common Core State Standards for Grade 3. Contributions to the Reading and Mathematics sections of this book were also made by the educational publishing staff at Show What You Know® Publishing. Dr. Jolie S. Brams, a clinical child and family psychologist, is the contributing author of the Worry Less About Tests and Test-Taking Hints for Test Heroes chapters of this book. Without the contributions of these people, this book would not be possible.

Table of Contents

Introduction v
Worry Less About Tests 1
Test-Taking Hints for Test Heroes 9
Reading 21
    Introduction 21
    Glossary of Reading Terms 22
    Reading Assessment One 25
        Directions 25
        Reading Assessment One Answer Document 51
    Reading Assessment Two 55
        Directions 55
        Reading Assessment Two Answer Document 81
Mathematics 85
    Introduction 85
    Glossary of Mathematics Terms 86
    Glossary of Mathematics Illustrations 94
    Mathematics Assessment One 99
        Directions 99
        Mathematics Assessment One Answer Document 119
    Mathematics Assessment Two 123
        Directions 123
        Mathematics Assessment Two Answer Document 145

Introduction

Dear Student:

This *Show What You Know® on the Common Core for Grade 3, Student Workbook* was created to give you lots of practice in preparation for your state proficiency test in Reading and Mathematics.

The first two chapters in this workbook—Worry Less About Tests and Test-Taking Hints for Test Heroes—were written especially for third-grade students. Worry Less About Tests offers advice on how to get rid of the bad feelings you may have about tests. The Test-Taking Hints for Test Heroes chapter gives you examples of the kinds of questions you will see on the test and includes helpful tips on how to answer these questions correctly so you can succeed on all tests.

The next two chapters of this Student Workbook help you prepare for the Reading and Mathematics test.
- The Reading chapter includes two full-length Reading Assessments and a Reading Glossary of words that will help you show what you know.
- The Mathematics chapter includes two full-length Mathematics Assessments, a Glossary of Mathematics Terms, and a Glossary of Mathematical Illustrations that will help you show what you know.

This Student Workbook will provide a chance to practice your test-taking skills to show what you know.

Good luck!

This page intentionally left blank.

Worry Less About Tests

Introduction

Many of us get nervous or anxious before taking a test. We want to do our best, and we worry that we might fail. You may have heard of your state's proficiency test, although you may not be familiar with the actual test. Because your state's proficiency test is new to you, you may become scared. You may worry about the test, and this might interfere with your ability to show what you know.

This chapter offers tips you can use on your state's proficiency test and many other tests. The ideas will build your test-taking confidence.

Worry Less About Tests

There are many things most of us would rather do than take a test. What would you rather do? Go to recess? See a movie? Eat a snack? Go swimming? Take a test? Most of us would not choose take a test. This doesn't mean we're afraid of tests. It means we like to do things that are more fun!

Some students do not want to take tests for another reason. They are afraid of tests and are afraid of failing. Even though they are smart enough to do well, they are scared. All of us worry about a test at one time or another. So, if you worry about tests, you are not alone.

When people worry about tests or are scared of tests, they have what is called test stress. You may have heard your parents say, "I'm feeling really stressed today." That means they have worried feelings. These feelings of stress can get in the way of doing your best. When you have test stress, it will be harder to show what you know. This chapter will help you get over your stress and worry less. You won't be scared. You will feel calm, happy, and proud.

If your mind is a mess
Because of terrible stress,
And you feel that you can't change at all.
Just pick up this book,
And take a look,
Our tips won't let you fall!

It's OK to Worry a Little Bit

Most people worry a little bit about something. Worrying isn't always a bad thing. A small amount of worrying is helpful. If you worry about crossing the street, you are more careful. When you worry about your school work, you work hard to do it right. As you can see, a little worrying isn't bad. However, you have to make sure you don't let worrying get in the way of doing your best. Think about crossing the street. If you worry too much, you'll never go anywhere. You can see how worrying too much is not a good thing.

What Kind of Kid Are You?

Test stress and worrying too much or too little can get in your way. The good news is there are ways you can help yourself do better on tests. All you have to do is change the way you think about taking tests. You can do better, not just by learning more or studying more, but by changing the way you think about things.

Now you will read about some students who changed the way they think about tests. You may see that these students have some of the same feelings you have. You will learn how each of these kids faced a problem and ended up doing better on tests.

Stay-Away Stephanie

Stephanie thought that it was better to stay away from tests than to try at all. She was scared to face tests. She thought, "If I stay home sick, I won't have to take the test. I don't care if I get in trouble; I'm just not going to take the test." Stay-Away Stephanie felt less nervous when she ran away from tests, but she never learned to face her fear. Stephanie's teacher thought Stephanie didn't care about tests or school, but this wasn't true at all. Stay-Away Stephanie really worried about tests. She stayed away instead of trying to face each challenge.

One day, Stephanie's mom had an idea! "Stephanie, do you remember when you were afraid to ride your bike after I took the training wheels off?" her mom said. "You would hide whenever I wanted to take a bike ride. You said, 'I would rather walk than learn to ride a two-wheel bike.' " Stephanie knew that wasn't true. She wanted to learn to ride her bike, but she was scared. She stayed away from the challenge. When Stephanie faced her fear, step by step, she learned to ride her bike. "Stephanie," her mom said, "I think you stay away from tests because you're worried." Stephanie knew her mom was right. She had to face tests step by step.

Stephanie and her teacher came up with a plan. First, Stephanie's teacher gave her two test questions to do in school. For homework, Stephanie did two more questions. When Stephanie was scared, she talked with her mom or her teacher. She didn't stay away. Soon, Stephanie knew how to ask for help, and she took tests without being worried. Now, she has a new nickname: Super-Successful Stephanie!

If you are like Stay-Away Stephanie, talk with your teacher or someone who can help you. Together, you can learn to take tests one step at a time. You will be a successful student instead of a stay-away student.

Worried Wendy

Wendy always thought that the worst would happen. Her mind worried about everything. "What if I can't answer all the questions? What if I don't do well? My teacher won't like me. My dad will be upset. I will have to study a lot more." Wendy spent her time worrying. Instead, she should have learned to do well on tests.

Wendy was so worried her stomach hurt. Wendy's doctor knew she wasn't sick; she was worried. "Wendy," he said, "I have known you ever since you were born. You have always been curious. You wanted to know how everything worked and where everything was. But now your curious mind is playing tricks on you. You are so worried, you're making yourself sick."

Wendy's doctor put a clock on his desk. "Look at this clock. Is it a good clock or a bad clock?" Wendy had no answer. "Believe it or not, Wendy, we can trick our minds into thinking it is good or bad. I'm going to say bad things about this clock as fast as I can. First, it's not very big. Also, because the clock is small, I might not read the time on it correctly. Since the clock is so small, I might lose it forever." Wendy agreed it was a bad clock. "But wait," said her doctor. "I think the clock is a neat shape, and I like the colors. I like having it in my office; it tells time well. It didn't cost much, so if I lose it, it isn't a big deal." Wendy realized she could look at tests the way the doctor looked at the clock. You don't have to worry. You can see good things, not bad.

Critical Carlos

Carlos always put himself down. He thought he failed at everything he did. If he got a B+ on his homework, he would say, "I made so many mistakes, I didn't get an A." He never said good things like, "I worked hard. I'm proud of my B+." Carlos didn't do well on tests because he told himself, "I don't do well on anything, especially tests."

Last week, Carlos got a 95% on a test about lakes and rivers. Carlos stared at his paper. He was upset. "What is the matter, Carlos?" his teacher asked. "Is something wrong?" Carlos replied, "I'm stupid; I missed five points. I should have gotten a 100%."

"Carlos, nobody's perfect: not me, not you, not anybody. I think 95 out of 100 is super! It's not perfect, but it is very good. Celebrate, Carlos!" Carlos smiled; he knew his teacher was right. Now, Carlos knows he has to feel good about what he does. He isn't sad about his mistakes. He's cheerful, not critical.

Look at the chart below. Use this chart to find out all the good things about yourself. Some examples are given to get you started.

Good Things About Me

1. *I make my grandmother happy when I tell her a joke.*

2. *I taught my dog how to shake hands.*

3. *I can do two somersaults in a row.*

4. I can tell her that I made someone happy.

5.

6.

Victim Vince

Vince couldn't take responsibility for himself. He said everything was someone else's fault. "Tests are too hard. I won't do well because they made the test too hard. And, last night, my little brother made so much noise that I couldn't write my homework story. It's his fault I won't do well. I asked Mom to buy my favorite snack. I have to have it when I study. She forgot to pick it up at the store. I can't study without my snack. It's her fault." Vince complained and complained.

Vince's aunt told him he had to stop blaming everyone for his troubles. "You can make a difference, Vince," she said. "When is your next test?" Vince told her he had a spelling test on Friday. "You're going to be the boss of the test. First, let's pick a time to study. How is every day at 4:00 p.m.?" Vince agreed. "Now, how are you going to study?"

"I like to practice writing the words a couple of times," Vince said. "Then, I ask Mom or Dad to quiz me."

"Great idea. Every day at 4:00 p.m., you're going to write each word four times. Then, ask one of your parents to review your list. You're the boss of the spelling test, Vince, because you have a plan."

Vince's Study Plan

Time	Monday	Tuesday	Wednesday	Thursday	Friday
					Spelling Test
4:00	Write down spelling words. Then, ask Mom or Dad to help.	Write down spelling words. Then, ask Mom or Dad to help.	Write down spelling words. Then, ask Mom or Dad to help.	Write down spelling words. Then, ask Mom or Dad to help.	
4:30					
5:00					
5:30					
6:00					
7:00	Look at spelling words again.	Look at spelling words again.	Look at spelling words again.	Look at spelling words again.	
7:30					Get a movie for doing well!

When Friday came, Vince's whole world changed. Instead of being in a bad mood because of a poor grade, Vince felt powerful! He took his spelling test and scored an A-. Vince could not believe his eyes! His teacher was thrilled. Vince soon learned he could control his attitude. Vince is no longer a victim. Instead, he is Victor Vince.

Perfect Pat

Pat spent all her time studying. She told herself, "I should study more. I should write this book report over. I should study every minute." Trying hard is fine, but Pat worked so much, she never felt she had done enough. Pat always thought she should be studying. Pat would play with her friends, but she never had a good time. In the middle of kickball or crafts, Pat thought, "I should be preparing for the test. I should be writing my book report." When Pat took a test, she worried about each question. "I can't answer this one. I should have studied harder."

"Pat," her principal said, "you have to relax. You're not enjoying school." Pat replied, "I can't do that. There is so much more to learn." The principal gave Pat some tips on how to use her study time better.

- Do not study for long periods of time. Instead, try to work for about 10–20 minutes at a time, and then take a break. Everyone needs a break!

- Ask yourself questions as you go along. After you study a fact, test yourself to see if you remember it. As you read, ask yourself questions about what you are reading. Think about what you are studying!

- Find a special time to study. You may want to think of a good time to study with the help of your parents or your teacher. You could choose to study from 4:30 to 5:00 every day after school. After dinner, you could complete homework from 7:30 to 8:00. After you finish studying, do not worry! You have done a lot for a third grader.

- Remember, you are a third-grade kid! School is very important, but playing, having fun, and being with your friends and family are also very important parts of growing up. Having fun does not mean you won't do well in school. Having fun in your life makes you a happier person and helps you do better on tests.

"Everyone Else Is Better" Edward

Edward worried about everyone else. During holidays, Edward thought about the presents other people received. At his baseball game, he worried his teammates would score more runs. Edward always wanted to know how his friends did on tests. He spent so much time worrying about what other people were doing, he forgot to pay attention to his own studying and test taking.

"Edward, you're not going to succeed if you don't worry about yourself," his grandfather told him. "You need to start talking about what you can do. Instead of asking your friend how he did on a test, you say, 'I got an 85%. Next time, I want to get a 90%.'" When Edward practiced this, he worried less about tests and was a whole lot happier.

Shaky Sam

Sam was great at sports. He was friendly and funny, and he had many friends. However, Sam had one big problem. Every time he thought about taking a test, he would start shaking inside. His heart would start pounding like a drum. His stomach would get upset. Even the night before a test, he started shaking really hard.

Sam's older brother liked to sing. He told Sam he used to get nervous before he sang to a crowd of people. "Sam, you need to trick your body. Don't think about the test; think about something fun and happy."

Sam closed his eyes. He thought about making four shots in a row on the basketball court. He thought about his favorite dessert: vanilla ice cream. He thought about swimming in his neighbor's pool. When he opened his eyes, he wasn't shaking.

Practice thinking happy thoughts, and make believe you are far away from your troubles. Test stress will disappear.

Other Ways That Third Graders Have Stopped Worrying About Tests

Third graders are pretty smart kids. They have lots of good ideas for getting rid of test stress. Here are some ideas from other third graders.

- When I am scared or worried, I talk to my neighbor. She is 70 years old. She is the smartest person I know. We sit on her porch and eat cookies and talk. It makes me feel better to know that she had some of the same problems when she was in third grade. She did well in school, and I know I can, too.

- Everything is harder in third grade, especially reading and math. I didn't want to go to school. I talked to my teacher, and he said we should have a talk every day before class. We talk about my homework, and he gives me tips. This really calms me down. When I am calm, I always do better.

- I used to worry that I wasn't doing well in school. I thought everyone else was smarter. My dad gave me a special folder. I keep all my tests in it. When I look at the tests, I see how much I have learned. I know I am doing a good job.

Kids are smart! You, your teachers, and your family and friends can help you find other ways to beat test stress. You will be surprised how much you know and how well you will do.

This page intentionally left blank.

Test-Taking Hints for Test Heroes

Introduction

Many third graders have not seen a test on which they have to fill in answer bubbles or write answers on lines. Before you sit down to take a test, it is a good idea to review problem-solving and test-taking hints.

This chapter offers many hints you can use when you take tests. The ideas will build your confidence and improve your test-taking skills.

Test-Taking Hints for Test Heroes   Show What You Know® on the Common Core for Grade 3

Do Your Best: Think Like a Genius!

Most third graders think the smartest kids do the best on tests. Smart kids may do well on tests, but all kids can do their best. By learning some helpful hints, most kids can do better than they ever thought they could on tests.

Learning to do well on tests will be helpful to you throughout your whole life, not just in third grade. Kids who are test-smart feel very good about themselves. They have an "I can do it" feeling about themselves. This feeling helps them succeed in school, in sports, and in music. It even helps with making friends. Test-smart kids usually do well in their school work, too. They believe they can do anything.

Become an Awesome Test Hero!

1. **Fill In the Answer Bubble**

   Think about tests you have taken. To answer questions, you may have written an answer, circled the correct answer, or solved a math problem. The state proficiency test is different. You will use your pencil to fill in answer bubbles. The test is mostly multiple choice, but there are a few short-answer questions for which you will write your answers on lines.

   For each multiple-choice question, you will have three choices to pick from. After you read the question and all the answer choices, think about which choice is correct. Next to each choice, you will see an answer bubble. The answer bubbles are not very big. They are smaller than the end of an eraser, smaller than a dime, and smaller than a jellybean. Even though the answer bubbles are small, they are very important. To answer each question, you must fill in the answer bubble next to the correct choice. Only fill in one answer bubble for each multiple-choice question. Fill in the bubble all the way, and do not color outside the bubble. Make sure you fill in the answer bubble neatly.

   Look at the example below. You can see the correct way to fill in an answer bubble. Practice filling in the answer bubbles in this example.

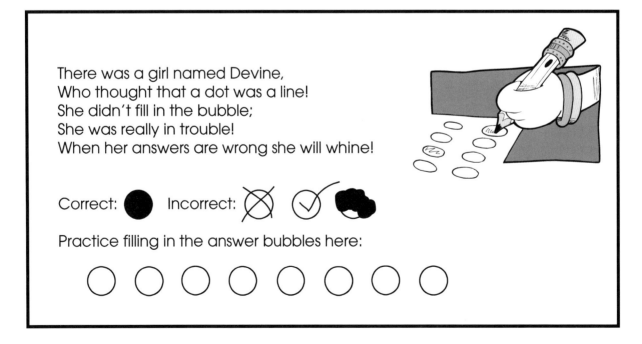

Learning how to fill in answer bubbles takes practice, practice, and more practice! It may not be how you are used to marking the correct answer, but it is one way to give a right answer. Think about Kay!

A stubborn girl named Kay,
Liked to answer questions her own way.
So her marked answer bubbles,
Gave her all sorts of troubles.
Her test scores ruined her day!

You will also have to answer short-answer questions. These are questions for which you have to write the answer. Some questions will only require one or two words or short phrases, but other questions may require a full sentence or two. Remember to write clearly and neatly so that people can read what you have written. Correct spelling and proper grammar will help to make your response clear. However, if you misspell a word or forget to use a comma or period, it will not be counted against you. The most important thing to remember when you answer short-answer questions is to completely answer the question as best you can.

2. **Only Fill In the Answer Bubbles You Need To**

It is not a good idea to touch the answer bubbles with your pencil until you are ready to fill in the right answer. If you put marks on more than one answer bubble, the computer that grades your test won't know which choice you think is right. Sometimes, kids get a little worried during the test. They might play with their pencils and tap their answer booklets. This is not a good idea. Look at all the answer choices. Only fill in one answer bubble for each multiple-choice question. This should be the answer bubble for the choice you think is right. Do not put marks in any other answer bubbles.

There was a nice girl named Sue,
Who thought she knew what to do.
She marked all the spots.
Her paper was covered with dots!
And she didn't show all that she knew.

3. Think Good Thoughts

The better you feel about taking tests, the better you will do. Imagine you are a famous sports hero. You feel good about playing your favorite sport. You feel good about yourself. As a sports hero, you don't start a soccer game, football game, baseball game, tennis match, or swimming meet by saying, "This is going to be hard. I can't do it." Instead you say, "This may be a little hard, but I can do it. I am glad I have a chance to do this. I am going to do my best. I know I can." You may think the state proficiency test is a little hard, but you can do it. When you start the state proficiency test, remember to think good thoughts. This will help you to be the best test hero you can be.

There was a girl named Gail,
Who thought she always would fail.
She said, "Tests are tough,
I'm not smart enough."
She had a sad end to her tale.

4. What Happens if I Don't Do Well on the Test?

The state proficiency test is one way to find out how much you have learned by the third grade. It is important to try your best on the state proficiency test, but remember, your friends, parents, and teachers will like you no matter how you do on this test.

There was a nice boy named Chad,
Who thought if he failed he was bad.
His teacher said, "That's not true.
I like you no matter how you do."
Now Chad is glad and not sad.

5. Don't Be Too Scared or Too Calm

Being too scared about tests will get in the way of doing your best. If you are scared, you won't be able to think clearly. If you are scared, your mind can't focus on the test. You think about other things. Your body might start to feel nervous. The chapter in this book called "Worry Less About Tests" will help you feel calmer about tests. Read that chapter so you can feel calmer about tests.

If you are too calm before taking a test, you might not do well. Sometimes, kids say, "I don't care about this!" They might not have pride in their school work. They may be nervous. They may think the state proficiency test is "no big deal" and may try to forget about it. If you do not think a test is important and you try to forget about it, you are not thinking good thoughts. Don't be scared of the test, but don't forget about it. You can become a test hero and do your best if you take pride in your work.

There was a student named Claire,
Who usually said, "I don't care."
Her sister named Bess,
Always felt total stress.
They weren't a successful pair!

6. Don't Rush; Speeding Through the Test Doesn't Help

The last time you took a test, did you look around the room to see who finished first? If someone handed his or her paper in before you, did you feel like you needed to hurry up? Kids feel that way sometimes, but rushing through questions will not help you. Finishing first, or second, or even third is not important. This may be a surprise to learn. Usually, we think speed is good. We hear about the fastest computer, the fastest runner, and the fastest car. Speed is exciting to think about, but working fast on your state proficiency test will not make your test score better. Take your time, and you will be able to show what you know!

There was a third grader named Liz,
Who sped through her test like a whiz.
She thought she should race
At a very fast pace,
But it caused her to mess up her quiz.

7. Read Directions Carefully!

One of the best ways to become a test hero is to read directions. Directions help you understand what you're supposed to do. It is really important to take your time and to read directions. You may say, "Why should I read directions? I know what to do." Here's a story that may change your mind.

Imagine you are a famous baker. Everyone thinks you make the best cakes in town! One day, a group of kings and queens comes for an important visit. They ask you to bake a special cake for them. You have never baked this type of cake before. The kings and queens give you directions, but you don't read them. You think to yourself, "Who has time? I don't need directions. I know how to bake cakes." You don't read the directions but put them in a drawer. This is not a good idea. The directions tell you to bake the special cake at 250 degrees, but you bake the cake at 350 degrees! What do you get? A very crispy cake and very angry kings and queens. You should have read the directions!

Make sure you read directions slowly and repeat them to yourself. You should understand the directions before you begin the test.

There was a nice boy named Fred,
Who ignored almost all that he read.
The directions were easy,
But he said, "I don't need these!"
He should have read them instead.

8. Don't Get Stuck on One Question

Some test questions will be easy. Other questions might be a little harder. Don't let that worry you! If there is a question you're not sure how to answer, use your pencil to put a mark by the question. Remember, mark the question, not the answer choice bubbles. Once you have marked the hard question, move on to the next question. When you get to the end of the test, go back and try to answer the hard question. Once you have answered many easy questions, you might be able to answer the hard question with no problem.

If you circle a question and move on, you won't get stuck. This is a good hint. Tests have lots of questions, so you will be able to show what you know. If there is a question that puzzles your mind, just go back to it later.

There was a sweet girl named Von,
Who got stuck and just couldn't go on.
She'd sit there and stare,
But the answer wasn't there.
Before she knew it, all the time was gone.

9. Use What You Know!

By the time you take your state's proficiency test, you will have been in school for four years. You went to kindergarten, first grade, second grade, and now you are in the third grade. You were taught lots of things in school, but you learned many things in other places, too. You may have gathered information at the library, in a magazine, from TV, from your parents, and from lots of other places. Third graders have a lot of information in their brains!

Sometimes, third graders forget how much they know. You may see a question that your teacher has not talked about. This is OK. You may have heard about it somewhere else. Take a minute to think about all you know.

Let's say you were asked the following question.

Melissa wants to walk to a store to buy an orange juice treat. The sign says the store is 200 yards away. If Melissa walks to the store, about how long will it take her?

A. About 5 minutes

B. About 30 minutes

C. About 1 hour

This seems like a hard question. You don't know how far 200 yards is. Stop and think for a minute! You have heard the word "yard" before, but where? You may have heard it used in a football game; a football field is 100 yards. So 200 yards is about the length of two football fields. You know it will not take long to walk that far. Now you know the right answer; it will probably only take about 5 minutes. Even though you thought you didn't know the answer, you used the information you remembered from other places. You're on your way to becoming a test hero!

There was a boy named Drew,
Who forgot to use what he knew.
He had lots of knowledge.
He could have been in college,
But his correct answers were very few.

10. Luck Isn't Enough!

Have you ever had a lucky number, a lucky color, or even a lucky hat? Everyone believes in luck. A famous football player always wears the same shoes game after game because he thinks they give him good luck. This doesn't make any sense. Wearing old, smelly shoes doesn't help him play well. But he believes in luck anyway. Believing in luck can be fun, but it is not going to help you do well on your state proficiency test. The best way to do well is to PRACTICE! Listening to your teacher, practicing the hints you have learned in this book, and learning every day in the third grade will help you do your best.

> There was a cool boy named Chuck,
> Who thought taking tests was just luck.
> He never prepared.
> He said, "I'm not scared."
> When his test scores appear, he should duck!

11. Recheck Your Answers

Everyone makes mistakes. Checking your work is very important. There once was a famous magician. He was very good at what he did, but he never checked his work. One night, he was getting ready for a big magic show. There were hundreds of people watching the show. The magician's wife said, "Check your pockets for everything you need." The magician didn't listen. "I've done this a million times," he said to himself. "I don't need to check my pockets." What a bad idea! When he got on stage, he reached his hand into an empty pocket—no magic tricks! Next time, he will recheck his pockets to do the best job possible! Going back and checking your work is very important. You can read a paragraph over again if there is something that you do not understand or something you forget. You will not be wasting time if you recheck your work. It is important to show what you know, not to show how fast you can go. Making sure you have put down the right answer is a good idea.

> There was a quick girl named Jen,
> Who read stuff once and never again.
> It would have been nice
> If she'd reread it twice.
> Her scores would have been better then!

Specific Strategies for Online Tests

Kids usually have two different kinds of thoughts about taking a test on a computer. Some say "Well, I use my computer all the time … I'm not going to even pay attention to the test … computers are easy!" Some kids think in the opposite way. They say "A computer test? That has to be even scarier than a regular test … there is no way I am going to do well!" The truth is that both of them are wrong. You have to use some special strategies to do your best on computer tests, and when you do, you will do your best!

1. **Read the Directions!** Here is a silly question: Would you want to eat a cake your friend made if he didn't read the directions on the box? Probably not! But even if you aren't a famous cook, you could make a pretty good cake if you read and follow directions. If you read the directions for EACH QUESTION you will have a much better chance of showing what you know. Because even if you know a lot, you have to answer what the question asks. Don't leave out this important step to test success!

2. **Don't Go With the First Answer.** Take a little time and read the WHOLE question and ALL the answer choices. The first answer that looks right is not always the best. Think about going out to dinner with your grandmother. You look at the menu and see "Big Ole Burger"! That sounds good. But if you looked at ALL the menu choices, you might have found your favorite tacos! The burger was good, but if you took more time, you would have found a better choice.

3. **Ask Yourself … How Much Time Do I Have?** You will have a certain amount of time to complete each section of the test. Always check to see how much time you will have. Practice also helps. Did you know that football players practice and practice to see how long it takes to line up and start a play? After a while they are more relaxed and don't worry about time running out. You need to take some practice tests to feel comfortable with timed tests.

4. **Is There a Good Way to Guess?** Most of the time it is a good idea to guess, especially if you can make an "educated" guess! That means you know some things about the question, but not everything. Let's say you aren't quite sure where your cousin lives, but you know it is cold and snows there all the time. One of your friends says that maybe your cousin lives in Georgia. You don't think that is right, because it hardly gets very cold there, and it is right next to Florida! So you can make an "educated guess" that "Georgia" isn't the right answer!

5. **When Should You Guess?** Unless the directions say that you will lose points for guessing, go for it! Educated guesses are the best, but even if you are really unsure of the answer, calm down and take a guess. If you have three possible answers, and make a guess, you have a one out of three chance of guessing correctly. That is like having two old pennies and one new penny in a bowl. If you just reach in, you will get the new penny one out of every three times you try. That's why you should answer every question!

6. **Don't Mess With That Test Window!** When people get a little nervous, they tend to make silly mistakes. One kid was rushing to make some toast before rushing off to school, and he unplugged the toaster instead of making the toast! Figure out how the computer screen works, and DON'T close that test window!

7. **Have a Good Attitude!** The better you feel, the better you will do! Remind yourself of how much you have learned in school. Remember that while this test is important, that your teachers will still like you a lot no matter how you do. Just do your best and feel good about yourself. Did you know that when runners have a good attitude, that they win more often? Well, the same goes for you and tests!

8. **If You Have Time Left, Use it!** You can use extra time to help you do your best! If your computer test allows, review your answers, especially if you guessed at a question or two. Take a deep breath and calm down. You might find that a better answer comes into your mind. Talk to yourself a little about some of your answers. You might ask yourself, "I chose the answer that said that it will take 6 hours for that ice cube to melt. That seems like a long time ... maybe I better recheck this and see if that makes sense."

Third graders all over have good ideas about tests. Here are some of them!

- Ask yourself, "Did I answer the question that was asked?" Carefully read the question so you can give the right answer.

- Read each answer choice before filling in an answer bubble. Sometimes, you read the first choice, and it seems right. But, when you get to the third choice, you realize that's the correct answer. If you had stopped with the first choice, you would have answered the question incorrectly. It is important to read all three choices before answering the question.

- Remember, nobody is trying to trick you. Do not look for trick answers. There will always be a right answer. If the answer choices do not look right, mark the question and go back to it later.

- Don't look around the room. Don't worry about how fast your friends are working, and don't worry about how well they are doing. Only worry about yourself. If you do that, you will do better on the test.

Reading

Introduction

In the Reading section of the *Show What You Know® on the Common Core for Grade 3, Student Workbook*, you will be asked questions to test what you have learned so far in school. These questions are based on the reading skills you have been taught in school through the third grade. The questions you will answer are not meant to confuse or trick you but are written so you have the best chance to show what you know.

The *Show What You Know® on the Common Core for Grade 3, Student Workbook* includes two full-length Reading Assessments that will help you practice your test-taking skills.

Glossary

alliteration: Repeating the same sound at the beginning of several words in a phrase or sentence. For example, "The bees buzzed in the back of the blue barn."

adjectives: Words that describe nouns.

adverbs: Words that describe verbs.

antonyms: Words that mean the opposite (e.g., *light* is an antonym of *dark*).

audience: The people who read a written piece or hear the piece being read.

author's purpose: The reason an author writes, such as to entertain, to inform, or to persuade.

author's tone: The attitude the writer takes toward an audience, a subject, or a character. Tone is shown through the writer's choice of words and details. Examples of tone are happy, sad, angry, gentle, etc.

base word (also called root word): The central part of a word that other word parts may be attached to.

biography: A true story about a person's life.

cause: The reason for an action, feeling, or response.

character: A person or an animal in a story, play, or other literary work.

compare: To use examples to show how things are alike.

contrast: To use examples to show how things are different.

details: Many small parts which help to tell a story.

descriptive text: To create a clear picture of a person, place, thing, or idea by using vivid words.

directions: An order or instructions on how to do something or how to act.

draw conclusion: To make a decision or form an opinion after considering the facts from the text.

effect: A result of a cause.

events: Things that happen.

fact: An actual happening or truth.

fiction: A passage that is made up rather than factually true. Examples of fiction are novels and short stories.

format: The way a published piece of writing looks, including the font, legibility, spacing, margins, and white space.

generalize: To come to a broad idea or rule about something after considering particular facts.

genres: Categories of literary and informational works (e.g., biography, mystery, historical fiction, poetry).

graphic organizer: Any illustration, chart, table, diagram, map, etc., used to help interpret information about the text.

heading: A word or group of words at the top or front of a piece of writing.

infer: To make a guess based on facts and observations.

inference: An important idea or conclusion drawn from reasoning rather than directly stated in the text.

inform: To give knowledge; to tell.

informational text (also called expository text): Text with the purpose of telling about details, facts, and information that is true (nonfiction). Informational text is found in textbooks, encyclopedias, biographies, and newspaper articles.

literary devices: Techniques used to convey an author's message or voice (e.g., figurative language, simile, metaphors, etc.).

literary text (also called narrative text): Text that describes actions or events, usually written as fiction. Examples are novels and short stories.

main idea: The main reason the passage was written; every passage has a main idea. Usually you can find the main idea in the topic sentence of the paragraph.

metaphor: A comparison between two unlike things without using the words "like" or "as." An example of a metaphor is, "My bedroom is a junkyard!"

Glossary

mood: The feeling or emotion the reader gets from a piece of writing.

nonfiction: A passage of writing that tells about real people, events, and places without changing any facts. Examples of nonfiction are an autobiography, a biography, an essay, a newspaper article, a magazine article, a personal diary, and a letter.

onomatopoeia: The use of words in which the sound of the word suggests the sound associated with it. For example, buzz, hiss, splat.

opinion: What one thinks about something or somebody; an opinion is not necessarily based on facts. Feelings and experiences usually help a person form an opinion.

passage: A passage or writing that may be fiction (literary/narrative) or nonfiction (informational/expository).

persuade: To cause to do something by using reason or argument; to cause to believe something.

plan: A method of doing something that has been thought out ahead of time.

plot: A series of events that make up a story. Plot tells "what happens" in a story, novel, or narrative poem.

plot sequence: The order of events in a story.

poetry: A type of writing that uses images and patterns to express feelings.

point of view: The way a story is told; it could be in first person, omniscient, or in third person.

predict: The ability of the reader to know or expect that something is going to happen in a text before it does.

prefix: A group of letters added to the beginning of a word. For example, *un*tie, *re*build, *pre*teen.

preposition: A word that links another word or group of words to other parts of the sentence. Examples are in, on, of, at, by, between, outside, etc.

problem: An issue or question in a text that needs to be answered.

published work: The final writing draft shared with the audience.

reliable: Sources used for writing that are trustworthy.

resource: A source of help or support.

rhyme: When words have the same last sound. For example, hat/cat, most/toast, ball/call.

root word (also called base word): The central part of a word that other word parts may be attached to.

schema: The accumulated knowledge that a person can draw from life experiences to help understand concepts, roles, emotions, and events.

sentence: A group of words that express a complete thought. It has a subject and a verb.

sequential order: The arrangement or ordering of information, content, or ideas (e.g., a story told in chronological order describes what happened first, then second, then third, etc.).

setting: The time and place of a story or play. The setting helps to create the mood in a story, such as inside a spooky house or inside a shopping mall during the holidays.

simile: A comparison between two unlike things, using the words "like" or "as." "Her eyes are as big as saucers" is an example of a simile.

solution: An answer to a problem.

stanzas: Lines of poetry grouped together.

story: An account of something that happened.

story elements: The important parts of the story, including characters, setting, plot, problem, and solution.

style: A way of writing that is individual to the writer, such as the writer's choice of words, phrases, and images.

suffix: A group of letters added to the end of a word. For example, teach*er*, color*ful*, sugar*less*, etc.

summary: To retell what happens in a story in a short way by telling the main ideas, not details.

Glossary

supporting details: Statements that often follow the main idea. Supporting details give you more information about the main idea.

symbolism: Something that represents something else. For example, a dove is a symbol for peace.

synonyms: Words with the same, or almost the same, meaning (e.g., sketch is a synonym of draw).

theme: The major idea or topic that the author reveals in a literary work. A theme is usually not stated directly in the work. Instead, the reader has to think about all the details of the work and then make an inference (an educated guess) about what they all mean.

title: A name of a book, film, play, piece of music, or other work of art.

tone: A way of writing that shows a feeling.

topic sentence: A sentence that states the main idea of the paragraph.

valid: Correct, acceptable.

verb: A word that shows action or being.

voice: To express a choice or opinion.

Reading Assessment One

Directions for Taking the Reading Assessment

The Reading Assessment contains seven reading selections and 40 questions. Some of the selections are fiction, while others are nonfiction. Read each selection and the questions that follow carefully. You may look back at any selection as many times as you would like. If you are unsure of a question, you can move to the next question and go back to the question you skipped later.

Multiple-choice questions require you to pick the best answer out of three possible choices. Only one answer is correct. The short-answer questions will ask you to write your answer and explain your thinking using words. Remember to read the questions and the answer choices carefully. You will mark your answers on the answer document.

When you finish, check your answers.

A Symbol of Pride

1. An important patriotic symbol of the United States is our national flag. Our flag represents the land, the people, and the government of the United States. On June 14, 1777, the Continental Congress declared (said) the flag of the United States would have thirteen stripes. The stripes would alternate red and white. This means one stripe would be red, the next stripe would be white; then, the next stripe would be red, and so on. The Continental Congress also said the flag would have a union (group) of thirteen white stars on a blue background. The thirteen stripes represented the thirteen original colonies. These thirteen colonies were the first states in the United States. The thirteen stars represented the number of states in the United States at that time.

2. The Continental Congress did not say how the stars should be arranged on the blue background, so flag makers used different designs. Sometimes, twelve stars were placed in a circle on the blue background with one star in the middle of the circle. Other times, the thirteen stars were placed in a circle on the blue background.

3. As new states became part of the United States, more stars and stripes were added to the flag. After awhile, people thought the flag had too many stripes. The Flag Act of 1818 stated the design of the American flag would include only thirteen stripes. The stripes were for the thirteen original colonies. The Flag Act also said the American flag should have one white star for each state that joins the United States. In 1846, the flag had 29 stars. By 1861, the number of stars had grown to 34, and in 1898, the flag contained 45 stars. The last change to the flag was in 1960 when a star was added for the state of Hawaii. On July 4 of that year, President Dwight D. Eisenhower approved the final arrangement of the 50 stars. The design of this 50-star flag is the one we still use today.

Go On

4   The American flag has had several nicknames over the years. Our country's earliest flag was known as the Continental flag or the Congress colors. Today, it is sometimes called the Stars and Stripes, Old Glory, or the Red, White, and Blue. No matter what name is used, the flag we see flying today is an important symbol of pride for our country.

The American Flag

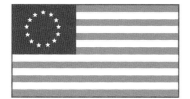  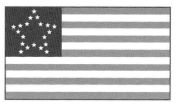

These are some examples of the American Flag. As you can see, there have been many different designs of the American Flag. Throughout the history of the United States, there have been more than 25 different designs. These are just three examples.

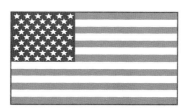

This is the current American Flag. The United States has used this design since 1960. The 50 stars represent our country's 50 states.

color key
= blue
= red
= white

Even though the American flag has had many designs, the colors have always been the same. Our flag includes red stripes, white stripes, white stars, and a blue square.

Go On

1. Read this sentence from the article.

   The Continental Congress did not say how the stars should be arranged on the blue background, so flag makers used different designs.

   Which word has the SAME meaning as the word *arranged*?

   A. makers
   B. say
   C. placed

2. In what year did President Eisenhower approve the design of the current American flag?

   A. 1777
   B. 1861
   C. 1960

3. Based on "The American Flag" section, how many stripes does the current American flag have?

   A. 14
   B. 13
   C. 50

4. How are the flag of 1818 and the current American flag ALIKE?

   A. The colors used on both flags are the same.
   B. The stripes on both flags represent cities in the United States.
   C. The stars on both flags represent the original colonies.

5. Why did the author MOST LIKELY write the article "A Symbol of Pride"?

   A. to tell about the history of the American flag

   B. to tell about President Dwight Eisenhower

   C. to explain the Flag Act of 1818

6. Who said the first flag of the United States would have thirteen stripes?

   A. President Eisenhower designed the flag with thirteen stripes.

   B. The Continental Congress said the flag would have thirteen stripes.

   C. The Flag Act of 1818 said the flag should have thirteen stripes.

Go On

Good Morning, Sunshine!

1   Annie woke up to a ray of sunlight on her face. She blinked her eyes and stretched her arms out from beneath her blankets. She couldn't believe it was morning already. It seemed as if she had just gone to sleep. Staring at the ceiling, she thought about her day and remembered what was going to happen in just two hours. Annie's stomach began to churn. She sat up, but fell back onto her pillow. She closed her eyes again, secretly wishing it would all be over.

2   Her mother knocked and opened the bedroom door just a crack. She saw that Annie was awake. "Good morning, Sunshine. It's time to get up! You don't want to waste a second this morning."

3   But that's exactly what Annie wanted to do. Slowly, Annie pulled herself out of bed and looked around the room that was now hers. It was still a mess. Moving boxes were everywhere. Four white, empty walls stared back at her. Maybe it would be better when things were unpacked. That's what her mom and dad kept telling her. Annie couldn't believe it would ever feel as comfortable as her old room.

4   The light of the sun made it possible to study each wall; it had been too dark the night before. One wall had a tiny crack that looked a little like a spider. "I'll never like a room with a spider crack," she thought to herself. Her mind was made up.

5   A pair of jeans, a purple shirt, and some sneakers—all brand new—were piled in the corner. Annie's mother surprised her daughter with the gift just yesterday. Annie knew her mother was trying to make her feel better, but Annie thought she might be more comfortable in her old clothes. New clothes never felt as good to her as her old ones. Since Annie didn't even know where to start looking for her old favorites, she settled for the new attire.

6   "Annie, I don't hear you moving. Are you up, Sunshine?"

7   "Why does she always call me that?" wondered Annie. It was a nickname Annie didn't like, but she had grown used to it. "She should call me 'Grumpy,'" Annie whispered to herself. "That's how I feel." With her shoes tied and her hair combed, she traveled into the unfamiliar hallway and down the stairs to greet a plate of pancakes.

8   "Oh, Annie, you look great!" her mom smiled as she poured a tall glass of milk.

9   "Thanks," Annie said, but she didn't really mean it. Around and around, she pushed syrup-soaked flapjacks around her plate. "How can I eat?" she thought. "My insides are tied in knots." Annie watched the clock count down the remaining moments of summer.

10   "Let's go, Annie." Her dad's voiced echoed. "You're going to miss the bus."

7. How does the picture of Annie let you know she is unhappy?

   A. It shows Annie's stuff is still in boxes.

   B. It shows Annie with a sad look on her face.

   C. It shows Annie cannot find her old clothes.

8. Why does Annie feel like her stomach is tied in knots?

   A. She is nervous about her first day at a new school.

   B. She doesn't like what she was going to have for breakfast.

   C. She has to wear her old clothes to school instead of the new ones.

9. Which word BEST describes Annie in this story?

   A. happy

   B. lonely

   C. grumpy

10. What is Annie's MAIN problem in the story?

    A. She hasn't cleaned her room for two weeks.

    B. She packed up all her things because she wants to paint the walls.

    C. She and her family just moved into a new house.

Go On

11. How is Annie's new room DIFFERENT from her old room?

   A. Annie's new room is more comfortable.

   B. Annie's new room has a spider crack.

   C. Annie's old room was bigger.

12. Read the sentence from the story.

   Annie's stomach began to churn.

   What is the author comparing Annie's stomach to in the above sentence?

   A. a spider hiding in her room

   B. a machine in which butter is made

   C. the pancakes that her mother is cooking

The Olympic Games

1   Today, the Olympic Games include some of the most popular sporting events in the world. Many of the events we see in today's Olympics are based on games from ancient Greece—games played over two thousand years ago. Believe it or not, many of the sports we watch today also took place a long time ago.

2   The first Olympic Games were held thousands of years ago. People would travel for many miles to gather for the five-day festival. The Games began with an opening ceremony. Visitors would watch men compete in footraces, wrestling, horse and chariot races, and the pentathlon. The pentathlon was a combination of five events. Athletes ran, jumped, threw a discus, threw a javelin, and wrestled. Only men were allowed to take part in the ancient Olympics. The Olympics were held every fours years in the summer. Only summer sports were part of the ancient Games. Winners were given a crown made of leaves. The events of the original Olympic Games were stopped about 1,600 years ago, around the year 393.

3   The first modern Olympic Games were held in 1896. These Games took place in Athens, Greece. About 300 athletes competed (played) in the Games. Only summer events were included. The tradition of the Olympic Games has continued. The Summer Olympic Games include some of the same events as the ancient games. Athletes throw the discus and javelin, run in races of different lengths, and wrestle. The modern Olympics have many more sports. These include swimming, diving, tennis, volleyball, and many more.

4   The modern Olympic Games are different from the ancient Olympics in other ways, too. The Games now offer more than summer sports. The first Winter Olympics were held in 1924. Some Winter Olympic sports are skiing, skating, hockey, and basketball. The Summer and Winter Games still begin with an opening ceremony, but each is held in a different city every four years. Also, the Olympic athletes of today include both men and women. Another difference: winners receive medals in gold for first place, silver for second place, and bronze for third place finishes.

5   While many things have changed, the Olympic Games continue to be an event where people watch their favorite athletes compete to be the best in their events. It is a time when people come together to cheer, to compete, and to show pride in their abilities and in their countries.

13. Read this sentence from the article.

    Many of the events we see in today's Olympics are based on games from ancient Greece.

    Ancient means that the games came from

    A. long ago
    B. yesterday
    C. modern times

14. How are the ancient Olympics and the modern Olympics ALIKE?

    A. Both included men and women.
    B. Both included summer and winter sports.
    C. Both awarded prizes to winners.

15. Why did the author MOST LIKELY write the selection?

    A. to inform the reader about the first Olympic Games in Greece
    B. to tell about the summer and winter Olympic Games
    C. to explain how the modern Olympics and the ancient Olympics are alike and different

16. When did the first modern Olympic games take place?

    A. 393
    B. 1924
    C. 1896

17. Read these sentences from the article.

    The pentathlon was a combination of five events. Athletes ran, jumped, threw a discus, threw a javelin, and wrestled.

    When an athlete competes in a *pentathlon*, they are competing in

    A. the opening ceremonies for the Olympic games.

    B. the race to win the gold, silver, or bronze medals.

    C. five athletic events in the summer Olympics.

18. In your own words, tell how are the modern Olympics DIFFERENT from the ancient Olympics?

Read this selection. Then answer the questions that follow.

Tony's Skunk

1   Tony had never been so scared. The light of his flashlight was fixed on a small animal. The bright eyes of a skunk stared at him. "You're smaller than my puppy," he whispered. He knew the animal wouldn't attack him. Yet, he was afraid to move a muscle. "What am I going to do?" he thought. Tony knew if he did the wrong thing, he would be very sorry.

2   This was the first time Tony had spotted a skunk. Tony didn't know much about the black and white creature, but he knew he must be careful. When skunks are scared, they spray a liquid that smells very bad. This liquid, which comes from a gland near the base of the tail, keeps predators away. Tony knew the skunk would growl and stomp its feet if it was about to spray. So, he carefully watched the little animal.

3   The skunk seemed startled. "He's as scared as I am," Tony thought. Tony's eyes were locked on the skunk. The skunk's eyes were fixed on Tony. A few more moments passed. Tony was ready to make a decision. "I hope I'm doing the right thing."

4   Tony began to move away slowly. He never let his eyes leave the skunk. He picked up his right foot, then his left. He softly whispered to himself, "I don't want to hurt you. Just let me get away from you."

5   The skunk was motionless. Tony moved away from him. Inch by inch, Tony backed away through the brush. He didn't want to move his flashlight off the skunk. He couldn't see where he was walking. He hoped he didn't stumble and fall. The leaves softly crunched under his feet. The skunk remained motionless.

6   He started to feel a bit of relief. Tony was about twelve feet from the little animal. He continued to step gently. Then suddenly, his foot came down and "CRACK!" The sound echoed about. It was only a dry stick, but it made Tony jump. The flashlight hit the ground. Light bounced around the brush. Tony's heart was beating fast. He couldn't see the skunk. The light no longer hit the skunk's eyes. Tony heard a small growl. "Oh no! I'm going to get it!" But Tony was lucky; the skunk had different plans. Tony heard four small footsteps running through the brush. Relieved, Tony scooped up his flashlight and quickly headed in the opposite direction.

19. Why was Tony scared?

    A. The light on his flashlight did not work.

    B. He thought the skunk might spray him.

    C. He was lost in the woods.

20. In your own words, tell what happens **after** Tony steps on the dry stick?

21. Read these sentences from the story.

    Light bounced around the brush. Tony's heart was beating fast.

    What mood does the author create by writing that *Tony's heart was beating fast*?

    A. scary

    B. calm

    C. funny

22. How does Tony try to solve his problem with the skunk?

    A. He throws his flashlight at the skunk.

    B. He cracks a stick to scare the skunk.

    C. He slowly steps away from the skunk.

23. How does the picture of Tony let you know he is scared?

    A. It shows Tony dropping his flashlight.

    B. It shows Tony with wide-open eyes.

    C. It shows Tony pointing the flashlight at the skunk.

Go On

Read this selection. Then answer the questions that follow.

1   There are two foods I never want to see on my plate: carrots and gravy. I don't mean together. I dislike them separately. For as long as I can remember, I have never liked these two. I'm not sure why. Mom says, even when I was little, carrots and gravy weren't for me.

2   We have a special rule in our house. At dinner, we have to try everything on our plates. I've tried carrots with salad and gravy with potatoes, but nothing works. I'm nine years old, and I still don't like carrots or gravy. I know I never will. A few weeks ago, carrots were served, again. I complained, but mom just looked at me. I knew the rule. I swallowed the orange mess. I grabbed my throat. I grabbed by belly. I fell from my chair. Mom just looked at me.

3   "I guess he's tried long enough. If Julio survives, no more carrots. I promise."

4   I lifted my head off the floor. "What about gravy?" I smiled.

5   "We'll see."

6   Mom kept her promise, and all was going great until a special family dinner. Mom and I met my grandparents at their favorite restaurant. We were celebrating their wedding anniversary. My uncle and his son Jake were there, too. Jake is a year older; he's in fourth grade. Jake and I both wanted hamburgers and french fries with lots of ketchup. We sat at the table, practicing our orders. A waitress came by and served a few glasses of water.

7   "Your dinners will be ready, soon."

8   "But we haven't ordered, yet." Jake and I were confused.

9   "It's been taken care of," she said. "I just know you're going to love it."

10   My uncle had called ahead to save time. I was disappointed. I was really looking forward to my hamburger. The waitress was right. Our dinner started in no time at all. The salad came first, but it was piled with little carrot slices. I tried to be polite. I pushed the orange pieces to the side of my plate. When my salad was free of carrots, I enjoyed every last bite. "This isn't so bad," I thought.

11   After taking away our salad plates, it was time for the main course. "Please!" I whispered to Jake, "let it be something I like!" I held my breath. The plate was in front of me. I peeked. Turkey and mashed potatoes covered in gravy! Mom smiled at me. I didn't grab my throat or my belly. Instead, I shrugged my shoulder and gave her half a smile. I spent most of the dinner just moving things around on my plate. Everyone was busy talking. No one seemed to notice.

12   "Are you going to eat that?" Jake asked. I shook my head no. He heaped the dinner onto his plate. "This stuff is great." I couldn't believe he thought so. All that gravy—yuck.

13   I thought my mom would be mad, but on our way home, she stopped at Happy Burger. I didn't even ask her to. "What will you have, Julio?" Mom asked.

14   "I'll take a cheeseburger and fries. No carrots. No gravy."

24. Why did Julio's mom stop at Happy Burger?

   A. She took Julio to Happy Burger to celebrate his birthday.

   B. She knew Julio didn't eat much at the restaurant.

   C. She was hungry.

25. What is Julio's MAIN problem in the story?

   A. He didn't like the restaurant he was going to with his family.

   B. He grabbed his throat and his belly and fell to the floor.

   C. He was served food he didn't like at the restaurant.

26. How are Julio and Jake ALIKE?

   A. They both wanted to order hamburgers.

   B. They both enjoyed the dinner at the restaurant.

   C. They are the same age.

27. Read this sentence from the story.

   He heaped the dinner onto his plate.

   When Jake *heaped* Julio's dinner onto his plate, he

   A. threw Julio's dinner at him.

   B. piled Julio's dinner on his plate.

   C. ate Julio's dinner.

28. In your own words, explain the MAIN idea of the story?

29. Why did the author MOST LIKELY write the story "Carrots and Gravy"?

   A. to convince readers to try all foods on their plates

   B. to tell a story about Julio, a boy who doesn't like carrots or gravy

   C. to explain why people should eat at Happy Burger instead of at nice restaurants

30. How do you know Julio's mom is not mad at him when he doesn't eat his dinner at the restaurant?

   A. She fixes him a hamburger when they get home from the restaurant.

   B. She lets him eat dessert, even though he didn't finish the main course.

   C. She stops at Happy Burger to get him some food.

Read this selection. Then answer the questions that follow.

1   Writing in a journal is a great way to keep track of things in your life. A journal can also be called a diary. Writers use journals to record their thoughts and ideas. Some people use journals to collect story ideas. Others write about bad days or happy times. Keeping a journal can help you reflect or think about what is happening in your life. When you read what you have written in your journal, you can learn from your everyday experiences.

2   It doesn't take any fancy materials to start a journal. You can start with a simple notebook and a pen or a pencil. You can get a special notebook if you want to make your journal more fancy. You can also use a computer to start your journal. The important thing is to record your thoughts and ideas. You can write them on paper or you can type them, whichever way is the best for you.

Go On

3   Try to write in your journal every day. You don't need to write for a long period of time, maybe only five or ten minutes, at first. As you keep practicing, you may find that you want to write for a few more minutes.

4   Writing a journal is a good way to practice your writing skills. Even though your journal is filled with your own words, try to use your best handwriting (if you're writing in a notebook). You want to be able to read what you have written! Also, try to spell words correctly and use complete sentences. Working on these skills will help you become a better writer.

5   Another secret to journal writing is to keep track of your writing. Put the date on each piece. This will help you remember when things happened. Every week or so, go back and read what you have written. You might find that something that seemed important has been forgotten. You might also find words that surprise you. Whatever the case, keep writing. You may be surprised at how much you have to say.

Journal Ideas

Write about things that happen each day.

Write about books you have read.

Keep a dialogue journal. You and another person (a friend, parent, or teacher) write back and forth to each other.

Write stories.

Write about something new you learn each day.

Journal Materials

computer

notebook and pencil

Important Things to Remember

Write every day.

June 9, 2011
Write the date on each piece.

Review what you have written.

Practice good writing skills.

Go On

31. Under which heading would you MOST LIKELY find information about items needed for journal writing?

    A. Journal Writing

    B. Journal Ideas

    C. Journal Materials

32. In your own words, explain the MAIN idea of this article?

33. Based on the article, how often does this author think you should write in your journal under the "Journal Writing" section?

    A. once a week

    B. twice a day

    C. once a day

34. Read the sentence from the article "Journal Writing."

    Another secret to journal writing is to keep track of your writing.

    A secret is

    A. something to tell everyone.

    B. something that is shared with few people.

    C. something that is unknown.

35. At the beginning of the article, what does the author say you can learn from reading what you have written in your journal?

    A. You can learn from your everyday experiences.

    B. You can learn new journal ideas.

    C. You can learn new words and phrases.

The Key

1   Emma noticed a small, white box on the stairs. She picked it up carefully and opened it to find a small key wrapped in a lace handkerchief. Faded pink and blue flowers decorated the delicate handkerchief, and the initials "A.B." were stitched into one corner. Emma assumed that it could belong to her grandmother because her name, Abby Brown, matched the initials.

2   As Emma examined the key, she saw that it was a dull, golden color. It was about as long as her index finger and seemed heavy for its size. There was a bit of crumpled blue ribbon looped through the hole at the top of the key. Emma thought to herself that the key must be very old because it was so worn. Emma wanted to find her grandmother and ask her about the key. As she went from room to room looking for her grandmother, Emma imagined that there might be wonderful treasures somewhere for that key to unlock. Then, Emma remembered that Gramma had gone to the store. Emma's questions would have to wait until she returned.

3   Emma was sitting on the porch step when Gramma returned. She slowly removed the key from her pocket and held it up for Gramma to see. "Look what I found—is it yours?" Emma asked. A broad smile appeared on Gramma's face as she inspected the key. "I thought I had lost this," Gramma said. "This key opens a very special box I have stored in the attic." Then, Gramma took Emma's hand and said, "Come with me. I have something to show you."

4   Emma and her grandmother walked through the house to the back stairs that led up to the attic. As they climbed the steep steps to the storage place, Emma's mind raced with thoughts about that special box. She pondered what it would look like and, most importantly, what would be inside the box.

36. Where was Emma waiting when her grandmother returned?

   A. inside the house

   B. in the backyard

   C. on the porch

37. In your own words, tell what happens AFTER Emma finds the key?

38. Read this sentence from the story.

   As Emma examined the key, she saw that it was a dull, golden color.

   Which word has the same base word as *golden*?

   A. goodness

   B. glider

   C. goldenly

39. How does the picture of Emma and her Gramma show you where the story takes place?

   A. It shows the stairs to the attic.

   B. It shows Emma finding the key.

   C. It shows Emma and her Gramma holding hands.

40. How are Emma and Gramma ALIKE?

   A. Both Emma and Gramma are nervous about what is inside of the special box.

   B. Both Emma and Gramma are excited about what is inside of the special box.

   C. Both Emma and Gramma are scared about what is inside of the special box.

1. Ⓐ Ⓑ Ⓒ
2. Ⓐ Ⓑ Ⓒ
3. Ⓐ Ⓑ Ⓒ
4. Ⓐ Ⓑ Ⓒ
5. Ⓐ Ⓑ Ⓒ
6. Ⓐ Ⓑ Ⓒ
7. Ⓐ Ⓑ Ⓒ
8. Ⓐ Ⓑ Ⓒ
9. Ⓐ Ⓑ Ⓒ
10. Ⓐ Ⓑ Ⓒ
11. Ⓐ Ⓑ Ⓒ
12. Ⓐ Ⓑ Ⓒ
13. Ⓐ Ⓑ Ⓒ
14. Ⓐ Ⓑ Ⓒ
15. Ⓐ Ⓑ Ⓒ
16. Ⓐ Ⓑ Ⓒ
17. Ⓐ Ⓑ Ⓒ

18. [blank response box]

19. Ⓐ Ⓑ Ⓒ

20. [blank response box]

21 Ⓐ Ⓑ Ⓒ
22 Ⓐ Ⓑ Ⓒ
23 Ⓐ Ⓑ Ⓒ
24 Ⓐ Ⓑ Ⓒ
25 Ⓐ Ⓑ Ⓒ
26 Ⓐ Ⓑ Ⓒ
27 Ⓐ Ⓑ Ⓒ

28

29 Ⓐ Ⓑ Ⓒ
30 Ⓐ Ⓑ Ⓒ
31 Ⓐ Ⓑ Ⓒ

32

33 (A) (B) (C)
34 (A) (B) (C)
35 (A) (B) (C)
36 (A) (B) (C)

37

38 (A) (B) (C)
39 (A) (B) (C)
40 (A) (B) (C)

Reading Assessment Two

Directions for Taking the Reading Assessment

The Reading Assessment contains seven reading selections and 40 questions. Some of the selections are fiction, while others are nonfiction. Read each selection and the questions that follow carefully. You may look back at any selection as many times as you would like. If you are unsure of a question, you can move to the next question, and go back to the question you skipped later.

Multiple-choice questions require you to pick the best answer out of three possible choices. Only one answer is correct. The short-answer questions will ask you to write your answer and explain your thinking using words. Remember to read the questions and the answer choices carefully. You will mark your answers on the answer document.

When you finish, check your answers.

MARSUPIALS

What are Marsupials?

1. The word marsupial means "pouched animal." Marsupials are mammals. They are warm-blooded animals. They are born live, and they drink their mothers' milk. There is something special about marsupials: they have a pouch. The female marsupial's pouch is a built-in nursery. Soon after its birth, a baby marsupial finds its way to its mother's pouch. A baby marsupial will live and grow in the pouch for several months.

2. Most marsupials live in Australia. Australia is a continent in the South Pacific Ocean. There are about 200 types of marsupials in the world. Types of marsupials include kangaroos, koalas, wombats, Tasmanian devils, wallabies, and opossums. The opossum is the only marsupial that lives in North America.

KANGAROOS

Facts about Kangaroos

3. Kangaroos usually come to mind when people think of pouched animals. There are many types of kangaroos, including the red, the rat, and the gray. Kangaroos grow to many sizes. Some are only one foot tall, but others grow to be six feet high. Kangaroos never run. Their back legs would get in the way if they tried. Instead, kangaroos use their long, strong back legs to help them jump up to three or four times their body's length. They also push off the ground with their tails as they begin a jump. Although kangaroos can jump at speeds of up to 30 miles per hour, they can only hop at this speed for very short distances.

4   Kangaroos live in groups called mobs. A mob may include 100 kangaroos. If you want to see a kangaroo, your best chance would be at night or very early in the morning when they are most active. During the middle of the day, they try to stay cool under trees or bushes. They live in many different settings (places) including grasslands, hills, or woodlands. Kangaroos like to graze on grass and other small plants.

Kangaroo Babies

5   Baby kangaroos weigh about one ounce at birth. They are about the size of a lima bean. They are blind at birth and have no fur. A baby kangaroo lives in its mother's pouch for about five months. Baby kangaroos are called joeys, but they are also known as "young-at-foot." Kangaroos usually give birth to only one baby at the time, but kangaroo mothers with "twins" have been spotted.

Facts about Koalas

6   The koala looks like a teddy bear, but it is not a bear at all. In fact, koalas may be more like monkeys. They live in trees, and they use their strong claws to hold onto branches in eucalyptus trees.

7   Koalas must live in thick eucalyptus forests, because adults need to eat between one and two pounds of the leaves daily. Koalas rarely drink water. They get the water they need from the leaves of the eucalyptus tree. The word "koala" actually means "no drink."

Baby Koalas

8   Koalas are one of the types of marsupials whose pouches open to the rear of their bodies. A baby koala lives in its mother's pouch for seven months. Then, the baby moves to the mother's back and lives in this "piggyback" position for five more months.

1. What is a kangaroo group called?

   A. a joey
   B. a mob
   C. a piggyback

2. In Paragraph 5, what word helps the reader know what *graze* means?

   A. different
   B. hills
   C. food

3. Which sentence BEST describes the main idea of this information?

   A. Koalas are like teddy bears.
   B. Kangaroos and koalas are two types of marsupials living in Australia.
   C. The opossum is the only marsupial that lives in North America.

Go On

4. Why do Koalas live in thick eucalyptus forests?

   A. because they have strong claws

   B. because they need to eat 1 to 2 pounds of leaves each day

   C. because they do not drink water

5. Why did the author include information about Kangaroos and Koalas in the same article?

   A. because they both live in Australia

   B. because they are both marsupials

   C. because they are both mammals

Birds of a Feather

1   It was a quiet Monday morning on Feathertree Lane. Spring had arrived. The race was on. Who would have the best nests this year? It was the only thing the birds of the neighborhood could think about. Mrs. Sparrow flew down the street. She watched the other birds from the neighborhood eagerly gathering sticks and grass for their nests.

2   Mrs. Sparrow landed at a bird feeder to have a snack. She picked up a bite of her favorite birdseed. Then, one of her neighbors, Mrs. Jay, landed next to her. "Have you talked to the new neighbors yet?" Mrs. Sparrow asked.

3   "No, I haven't, but I can see they've been busy. Have you seen the nest they're building? It's beautiful," Mrs. Jay replied.

4   "Yes, it seems to be quite a nest. I wonder where they got the idea. I've never seen a bird who could weave grasses together in the way that they have."

5   Mrs. Jay wiped some birdseed from her beak before she answered. "They do seem to be good at constructing nests. I've been watching. They are so close to my nest. Their nest might be the biggest one in our neighborhood. But I'm quite happy in our nest. My husband, Blue, did a wonderful job finding the right grasses, sticks, bark, and feathers for our nest this year. It may not be the largest, but it is just right for our family. I'm very proud of our home."

6   "As you should be," said Mrs. Sparrow. "It does look quite cozy. Do you remember two years ago when the Hummingbird family made their nest here? It was so tiny. It was even smaller than my home."

7   "That's true," Mrs. Jay chirped. "They were always humming while they worked. I wondered if they had forgotten the words to their songs."

8   Mrs. Sparrow laughed. "No, no, no. The humming came from their wings moving so fast. But I never thought they were very friendly. And the Robins! They're not so nice! Just because they have those beautiful red feathers doesn't mean they should be mean. Look at the Cardinals. They have beautiful red feathers, too. They don't fly away when another bird comes by the feeder. They'll talk to any bird in the neighborhood. Besides, I think we're all beautiful in our own ways. Don't you?"

9   "Yes, I do agree with you. Oh, no! The people in the yellow house let their dog out. When he sees us, we won't be able to hear ourselves chirp. I think I'll head back to my nest rather than listen to him. I'll see you later, Mrs. Sparrow." Mrs. Jay chirped her goodbye and flew away.

10   Mrs. Sparrow ate another seed before she heard the barking begin. "There goes the neighborhood," she thought to herself as she flew home.

6. Read this sentence from the first paragraph of the story.

   She watched the other birds from the neighborhood eagerly gathering sticks and grass for their nests.

   What does the word *eagerly* mean?

   A. excited
   B. bored
   C. tired

7. In your own words, tell why Mrs. Jay flies away from the bird feeder.

8. What is this story mostly about?

   A. which birds can build the best nest for spring
   B. when the neighborhood dog gets out
   C. when Mrs. Jay and Mrs. Sparrow visit at the bird feeder

9. You can tell Mrs. Sparrow thinks the Robins are mean because they

   A. fly away when other birds come to the feeder.
   B. have red feathers.
   C. talk to the other birds.

Go On

10. When did the hummingbird family make their nest in the neighborhood?

    A. one year ago

    B. two years ago

    C. last Spring

11. Read the sentences from the fifth paragraph of the story.

    They do seem to be good at constructing nests. I've been watching.

    What does the word *constructing* mean?

    A. criticizing

    B. building

    C. destroying

12. From the story, the reader can tell that

    A. nest building happens every spring.

    B. neighborhood birds are all friendly.

    C. Mrs. Jay doesn't have her own nest.

Go On

The Harrison Elementary Press
A Newspaper Written By Kids, For Kids

March Issue   Science Section, Page 1

FROGS AND TOADS
By Federico Garcia

It's important to look to see if the animal you are about to kiss is a frog or a toad. You may never find a handsome prince if you kiss the wrong amphibian. Can you tell the difference between a frog and a toad?

It is easy to confuse frogs and toads just by looking at them. They are both amphibians. This means they can live both in water and on land. They both are coldblooded. This means their body temperatures are the same as the air temperatures around them. They have to look for cool, shady places to rest if they become too hot. Frogs and toads look for warm, sunny places if they are too cold. Both animals are vertebrates. This means they have spines. Their body shape is almost the same. Their eyes protrude out from their faces, so they can see in most directions without turning their heads.

Frogs and toads use their long, sticky tongues to catch insects to eat. Both frogs and toads swallow their food whole.

How are frogs and toads different? Frogs are better swimmers and jumpers because they have long back legs. A toad's back legs are shorter. Frogs are more likely to be found near water. Toads often live in drier places. Most frogs have four webbed feet. Toads do not have webs on their back feet. The skin of a frog is smooth and damp. Toads have drier skin that is covered with bumps called glands. Frogs have teeth in their upper jaws and no teeth in their lower jaws. Toads have no teeth at all.

As you can see, frogs and toads are not the same type of amphibian. Of course, a frog turning into a handsome prince only happens in fairy tales. Who would kiss a frog or a toad anyway?

Go On

13. Look at the chart below.

Differences in	
Frogs	**Toads**
• Long back legs	• Shorter back legs
• Found near water	
• Smooth and damp	• Dry bumpy skin
• Four webbed feet	• No webs on back feet

Which item belongs in the blank?

A. live in drier places

B. can turn into a handsome prince

C. have smooth damp skin

14. Read this sentence from the second paragraph of the article.

Their eyes protrude out from their faces, so they can see in most directions without turning their heads.

What does the word *protrude* mean?

A. sink in

B. stick out

C. lay flat

15. What is the MAIN idea of this article?

   A. how to find a handsome prince

   B. how frogs and toads catch their food

   C. how frogs and toads are similar and different

16. Look at the Venn diagram below.

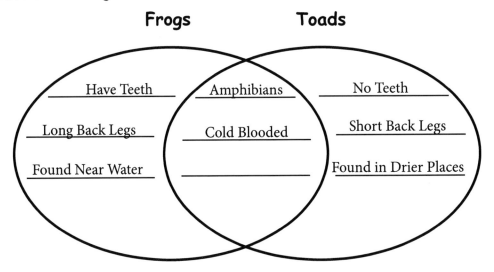

Which item belongs in the blank?

   A. Vertabrates

   B. Dry bumpy skin

   C. Four webbed feet

17. According to the article, what do frogs and toads have to do if the temperature becomes too cold?

   A. They begin to hibernate.

   B. They find a place to rest.

   C. They find a warmer place.

18. In your own words, write a summary of Federico's newspaper article on frogs and toads.

Read this selection. Then answer the questions that follow.

The Raindrops' New Dresses

1. "We're so tired of these gray dresses!"
   Cried the little drops of rain,
   As they came down helter-skelter[1]
   From the Nimbus cloud fast train.

2. And they bobbed against each other
   In a spiteful[2] sort of way,
   Just like children when bad temper
   Gets the upper hand some day.

3. Then the Sun peeped out a minute.
   "Dears, be good and do not fight,
   I have ordered you new dresses,
   Dainty robes of purest white."

4. Ah! Then all the tiny raindrops
   Hummed a merry glad refrain[3],
   And the old folks cried: "How pleasant
   Is the music of the rain!"

5. Just at even, when the children
   Had been safely tucked in bed,
   There was such a rush and bustle
   In the dark clouds overhead!

6. Then those raindrops hurried earthward,
   At the North Wind's call, you know,
   And the wee folks, in the morning,
   Laughed to see the flakes of snow.

---

[1] Helter-skelter: with hurry and confusion

[2] Spiteful: having or showing a desire to cause harm or pain to others

[3] Refrain: a melody or tune

Go On

19. Read this line from the fourth stanza of the poem.

    Hummed a merry glad refrain[3]

    What does the word *refrain* mean?

    A. word
    B. poem
    C. melody

20. What can the reader tell from the poem?

    A. The raindrops didn't get their wish.
    B. The raindrops turned into snowflakes.
    C. The North Wind didn't blow.

21. In the beginning of the poem, the raindrops feel

    A. joyful.
    B. angry.
    C. unhappy.

22. What happened before the sun peeped out?

    A. The raindrops were in a bad mood.
    B. The clouds turned gray.
    C. The raindrops hummed a song.

23. In your own words, finish the summary of the poem.

    One day the raindrops became tired of their gray dresses and acted badly.

Clouds

1. Clouds are little drops of water or ice that float together through the air. Clouds come in different shapes and sizes.

2. There are three different kinds of clouds. You can tell the kind of cloud by the way the cloud looks and where it is in the sky. Some clouds are low in the sky, some are in the middle, and some are high in the sky.

3. The highest clouds in the sky are cirrus clouds. Some people think cirrus clouds look like thin white feathers. Others call cirrus clouds "mares' tails" because they look like the long tails of horses. Because cirrus clouds are so high in the sky where the temperature is very cold, cirrus clouds are made of tiny ice crystals. Cirrus clouds move quickly across the sky at about 200 miles an hour.

4. Another kind of cloud is the cumulus cloud. These clouds look like pretty white cotton balls in the sky. Cumulus clouds hover low in the sky and change shape as the air moves them. Some people like to play games to guess what shape the cumulus cloud looks like: a ship, a flower, a face, or an animal. When cumulus clouds pile on top of each other they are called nimbus clouds. Nimbus clouds turn into black storm clouds that bring thunderstorms.

5. Stratus clouds are low in the sky. They look like long, gray blanket clouds. Rain and snow may fall from stratus clouds. Stratus clouds often hide the sun and the moon.

Go On

24. What are paragraphs 3, 4, and 5 mostly about?

   A. people playing games with clouds

   B. the different types of clouds

   C. the shapes of clouds

25. Read this sentence from the fourth paragraph of the story.

   Cumulus clouds hover low in the sky and change shape as the air moves them.

   What does the word *hover* mean?

   A. float

   B. glide

   C. wait

26. The type of cloud that produces thunderstorms is called

   A. cirrus.

   B. cumulus.

   C. nimbus.

27. Look at the chart below.

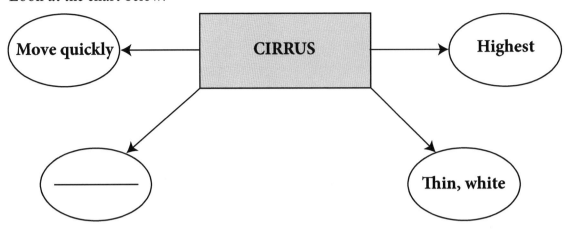

From paragraph 3, which detail about cirrus clouds belongs in the empty box?

A. Long, gray

B. Lowest in the sky

C. Made of ice crystals

28. Which of the following is the best summary of the passage?

A. There are different types of clouds. The highest ones are called cirrus clouds. Cumulus is another type of cloud. Not all clouds produce rain. Clouds have different shapes and sizes.

B. Clouds are formed from drops of water or ice floating together. They come in different shapes and sizes and can be high or low in the air. Some clouds move quickly, while others produce rain.

C. Cumulus is a white, fluffy cloud that looks like cotton balls in the air. People play games guessing what shape they look like. Nimbus clouds produce thunderstorms and are dark and gray.

Read this selection. Then answer the questions that follow.

The Farmer and His Three Sons

1   A farmer who had worked hard all his life was taken sick. He knew that he must soon die. He called his three sons about his bed to give them some advice.

2   "My sons," said he, "keep all of the land which I leave you. Do not sell any of it, for there is a treasure in the soil. I shall not tell you where to hunt for it, but if you try hard to find it, and do not give up, you will surely prosper."

3   "As soon as the harvest[1] is over, begin your search with plow, and spade, and rake. Turn every foot of earth, then turn it again and again. The treasure is there."

4   After the father died, the sons gathered in the harvest. As soon as the grain had been cared for, they planned to search for the hidden treasure. The farm was divided into three equal parts. Each son agreed to dig carefully his part.

5   Every foot of soil was turned by the plow or by the spade[2]. It was next harrowed[3] and raked, but no treasure was found. That seemed very strange.

6   "Father was an honest man and a wise man," said the youngest son. "He would never have told us to hunt for the treasure if it were not here. Do you not remember that he said, 'Turn the soil again and again'? He surely thought the treasure worth hunting for."

7   "Our land is in such good condition now that we might as well sow winter wheat," said the oldest son. His brothers agreed to this and the wheat was sown.

[1] Harvest: the gathering in of a crop
[2] Spade: a tool used for digging
[3] Harrow(ed): to break up and level plowed ground

8   The next harvest was so great that it surprised them. No neighbor's field bore[4] so many bushels[5] of wheat to the acre. The sons were pleased with their success.

9   After the wheat was harvested, they met to make plans for searching again for the hidden treasure. The second son said:

10   "I have been thinking ever since our big harvest that perhaps father knew how this search would turn out. We have much gold. We did not find it in a hole in the ground, but we found it by digging. If we had not cultivated[6] our fields well, we should not have had such a crop of wheat. Our father was wise; we have dug for the treasure and have found it.

11   "We will cultivate the ground still better next year and make the soil rich; then we shall find more treasure."

12   The other sons agreed to this. "It is good to work for what we get," they said.

13   Year after year the farm was well tilled and bore good crops. The sons became rich, and they had two things much better than wealth—good health and happiness.

[4] Bore: produced
[5] Bushel: a unit of dry measure for grain
[6] Cultivated: to have prepared soil or land for growing crops

Go On

29. Read this sentence from the second paragraph of the story.

I shall not tell you where to hunt for it, but if you try hard to find it, and do not give up, you will surely prosper.

What does the word *prosper* mean?

A. fail

B. succeed

C. quit

30. In paragraph 3, as soon as the harvest is over, what did the father tell his sons to do first?

A. plow

B. spade

C. dig

31. The farmer gave his sons advice because he

A. was retiring.

B. was very ill.

C. wanted them to learn.

32. What can the reader tell from the sons' actions?

A. The sons didn't listen well.

B. The sons gave up searching for the treasure.

C. The sons respected their father.

33. In your own words, tell what the father meant by "treasure." Use details from the story to support your answer.

The Little Pine Tree

1 A little pine tree was in the woods.

2 It had no leaves. It had needles.

3 The little tree said, "I do not like needles. All the other trees in the woods have pretty leaves. I want leaves, too. But I will have better leaves. I want gold leaves."

4 Night came and the little tree went to sleep. A fairy came by and gave it gold leaves.

5 When the little tree woke it had leaves of gold.

6 It said, "Oh, I am so pretty! No other tree has gold leaves."

7 Night came.

8 A man came by with a bag. He saw the gold leaves. He took them all and put them into his bag.

9 The poor little tree cried, "I do not want gold leaves again. I will have glass leaves."

10 So the little tree went to sleep. The fairy came by and put the glass leaves on it.

11 The little tree woke and saw its glass leaves.

12 How pretty they looked in the sunshine! No other tree was so bright.

13 Then a wind came up. It blew and blew.

14 The glass leaves all fell from the tree and were broken.

15   Again the little tree had no leaves. It was very sad and said, "I will not have gold leaves, and I will not have glass leaves. I want green leaves. I want to be like the other trees."

16   And the little tree went to sleep. When it woke, it was like other trees. It had green leaves.

17   A goat came by. He saw the green leaves on the little tree. The goat was hungry and he ate all the leaves.

18   Then the little tree said, "I do not want any leaves. I will not have green leaves, nor glass leaves, nor gold leaves. I like my needles best."

19   And the little tree went to sleep. The fairy gave it what it wanted.

20   When it woke, it had its needles again. Then the little pine tree was happy.

34. According to the story, which event happens first?

   A. The little tree asks for gold leaves.

   B. The glass leaves fall from the tree.

   C. The goat eats all the leaves.

35. Based on the story, how does the little pine tree feel about the other trees in the woods?

   A. The little pine tree is mad at the other trees.

   B. The little pine tree is jealous of the other trees.

   C. The little pine tree is happy for the other trees.

36. In your own words, tell about the problem the little pine tree faces?

37. According to the story, what happened when the wind blew and blew?

   A. A man came by and put the gold leaves in his bag.

   B. A goat ate all the leaves.

   C. The glass leaves fell from the tree and were broken.

Go On

38. What can you tell from the story?

   A. The little tree was the only pine tree in the woods.

   B. The little tree was happy being a pine tree.

   C. There were no other trees in the woods.

39. What is the MAIN idea of the story?

   A. The little pine tree was alone in the woods.

   B. The little pine tree learns to be happy with its needles.

   C. Fairies change the leaves of trees.

40. How does the picture of the little pine tree let you know it is different?

   A. It shows the little pine standing off by itself.

   B. It shows the little pine tree as the only tree with needles.

   C. It shows the little pine tree with its gold leaves.

This page intentionally left blank.

Show What You Know® on the Common Core for Grade 3
Reading Assessment Two
Answer Document

1. Ⓐ Ⓑ Ⓒ
2. Ⓐ Ⓑ Ⓒ
3. Ⓐ Ⓑ Ⓒ
4. Ⓐ Ⓑ Ⓒ
5. Ⓐ Ⓑ Ⓒ
6. Ⓐ Ⓑ Ⓒ

7. [   ]

8. Ⓐ Ⓑ Ⓒ
9. Ⓐ Ⓑ Ⓒ
10. Ⓐ Ⓑ Ⓒ
11. Ⓐ Ⓑ Ⓒ
12. Ⓐ Ⓑ Ⓒ
13. Ⓐ Ⓑ Ⓒ
14. Ⓐ Ⓑ Ⓒ
15. Ⓐ Ⓑ Ⓒ
16. Ⓐ Ⓑ Ⓒ
17. Ⓐ Ⓑ Ⓒ

18.

19. Ⓐ Ⓑ Ⓒ
20. Ⓐ Ⓑ Ⓒ
21. Ⓐ Ⓑ Ⓒ
22. Ⓐ Ⓑ Ⓒ

23.

24 Ⓐ Ⓑ Ⓒ
25 Ⓐ Ⓑ Ⓒ
26 Ⓐ Ⓑ Ⓒ
27 Ⓐ Ⓑ Ⓒ
28 Ⓐ Ⓑ Ⓒ
29 Ⓐ Ⓑ Ⓒ
30 Ⓐ Ⓑ Ⓒ
31 Ⓐ Ⓑ Ⓒ
32 Ⓐ Ⓑ Ⓒ

33

Reading Assessment Two
Answer Document

34 Ⓐ Ⓑ Ⓒ
35 Ⓐ Ⓑ Ⓒ

36

[   ]

37 Ⓐ Ⓑ Ⓒ
38 Ⓐ Ⓑ Ⓒ
39 Ⓐ Ⓑ Ⓒ
40 Ⓐ Ⓑ Ⓒ

Mathematics

Introduction

In the Mathematics section of the *Show What You Know® on the Common Core for Grade 3, Student Workbook*, you will be asked questions to test what you have learned so far in school. These questions are based on the mathematics skills you have been taught in school through the third grade. The questions you will answer are not meant to confuse or trick you but are written so you have the best chance to show what you know.

The *Show What You Know® on the Common Core for Grade 3, Student Workbook,* includes two full-length Mathematics Assessments that will help you practice your test-taking skills.

Glossary

addend: Numbers added together to give a sum. For example, 2 + 7 = 9. The numbers 2 and 7 are addends.

addition: An operation joining two or more sets where the result is the whole.

a.m.: The hours from midnight to noon; from Latin words *ante meridiem* meaning "before noon."

analyze: To break down information into parts so that it may be more easily understood.

angle: A figure formed by two rays that meet at the same end point called a vertex. Angles can be obtuse, acute, right, or straight.

area: The number of square units needed to cover a region. The most common abbreviation for area is A.

Associative Property of Addition: The grouping of addends can be changed and the sum will be the same.
Example: (3 + 1) + 2 = 6; 3 + (1 + 2) = 6.

Associative Property of Multiplication: The grouping of factors can be changed and the product will be the same.
Example: (3 x 2) x 4 = 24; 3 x (2 x 4) = 24.

attribute: A characteristic or distinctive feature.

average: A number found by adding two or more quantities together and then dividing the sum by the number of quantities. For example, in the set {9, 5, 4}, the average is 6: 9 + 5 + 4 = 18; 18 ÷ 3 = 6. *See mean.*

axes: Plural of axis. Perpendicular lines used as reference lines in a coordinate system or graph; traditionally, the horizontal axis (*x*-axis) represents the independent variable and the vertical axis (*y*-axis) represents the dependent variable.

bar graph: A graph using bars to show data.

capacity: The amount an object holds when filled.

chart: A way to show information, such as in a graph or table.

circle: A closed, curved line made up of points that are all the same distance from a point inside called the center.
Example: A circle with center point *P* is shown below.

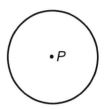

circle graph: Sometimes called a pie chart; a way of representing data that shows the fractional part or percentage of an overall set as an appropriately-sized wedge of a circle.
Example:

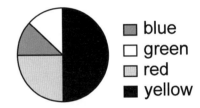

circumference: The boundary line or perimeter of a circle; also, the length of the perimeter of a circle.
Example:

Commutative Property of Addition: Numbers can be added in any order and the sum will be the same.
Example: 3 + 4 = 4 + 3.

Commutative Property of Multiplication: Numbers can be multiplied in any order and the product will be the same.
Example: 3 x 6 = 6 x 3.

compare: To look for similarities and differences. For example, is one number greater than, less than, or equal to another number?

conclusion: A statement that follows logically from other facts.

Glossary

cone: A solid figure with a circle as its base and a curved surface that meets at a point.

cones

congruent figures: Figures that have the same shape and size.

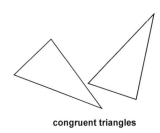
congruent triangles

cube: A solid figure with six faces that are congruent (equal) squares.

cylinder: A solid figure with two circular bases that are congruent (equal) and parallel to each other connected by a curved lateral surface.

data: Information that is collected.

decimal number: A number expressed in base 10, such as 39,456, where each digit's value is determined by multiplying it by some power of 10.

denominator: The bottom number in a fraction.

diagram: A drawing that represents a mathematical situation.

difference: The answer when subtracting two numbers.

distance: The amount of space between two points.

dividend: A number in a division problem that is divided. Dividend ÷ divisor = quotient. Example: In 15 ÷ 3 = 5, 15 is the dividend.

$$\text{divisor} \overline{)\text{dividend}}^{\text{quotient}} \qquad 3\overline{)15}^{5}$$

divisible: A number that can be divided by another number without leaving a remainder. Example: 12 is divisible by 3 because 12 ÷ 3 is an integer, namely 4.

division: An operation that tells how many equal groups there are or how many are in each group.

divisor: The number by which another number is divided. Example: In 15 ÷ 3 = 5, 3 is the divisor.

$$\text{divisor} \overline{)\text{dividend}}^{\text{quotient}} \qquad 3\overline{)15}^{5}$$

edge: The line segment where two faces of a solid figure meet.

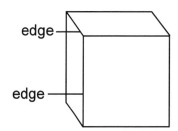

equivalent fractions: Two fractions with equal values.

equality: Two or more sets of values that are equal.

equation: A number sentence that says two expressions are equal (=). Example: 4 + 8 = 6 + 6.

estimate: To find an approximate value or measurement of something without exact calculation.

even number: A whole number that has a 0, 2, 4, 6, or 8 in the ones place. A number that is a multiple of 2. Examples: 0, 4, and 678 are even numbers.

expanded form: A number written as the sum of the values of its digits. Example: 546 = 500 + 40 + 6.

expression: A combination of variables, numbers, and symbols that represent a mathematical relationship.

Glossary

face: The sides of a solid figure. For example, a cube has six faces that are all squares. The pyramid below has five faces—four triangles and one square.

fact family: A group of related facts using the same numbers.
Example: 5 + 8 = 13; 13 − 8 = 5.

factor: One of two or more numbers that are multiplied together to give a product.
Example: In 4 × 3 = 12, 4 and 3 are factors of 12.

figure: A geometric figure is a set of points and/or lines in 2 or 3 dimensions.

flip (reflection): The change in a position of a figure that is the result of picking it up and turning it over.
Example: Reversing a "b" to a "d." Tipping a "p" to a "b" or a "b" to a "p" as shown below:

fraction: A symbol, such as $\frac{2}{8}$ or $\frac{5}{3}$, used to name a part of a whole, a part of a set, or a location on the number line.
Examples:

$$\frac{\text{numerator}}{\text{denominator}} = \frac{\text{dividend}}{\text{divisor}}$$

$$\frac{\text{\# of parts under consideration}}{\text{\# of parts in a set}}$$

function machine: Applies a function rule to a set of numbers, which determines a corresponding set of numbers.
Example: Input 9 → Rule × 7 → Output 63. If you apply the function rule "multiply by 7" to the values 5, 7, and 9, the corresponding values are:

5 → 35
7 → 49
9 → 63

graph: A "picture" showing how certain facts are related to each other or how they compare to one another. Some examples of types of graphs are line graphs, pie charts, bar graphs, scatterplots, and pictographs.

grid: A pattern of regularly spaced horizontal and vertical lines on a plane that can be used to locate points and graph equations.

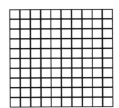

hexagon: A six-sided polygon. The total measure of the angles within a hexagon is 720°.

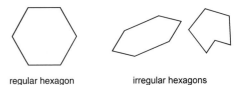

regular hexagon   irregular hexagons

impossible event: An event that can never happen.

integer: Any number, positive or negative, that is a whole number distance away from zero on a number line, in addition to zero. Specifically, an integer is any number in the set {. . .-3,-2,-1, 0, 1, 2, 3. . .}.
Examples of integers include: 1, 5, 273, -2, -35, and -1,375.

intersecting lines: Lines that cross at a point.
Examples:

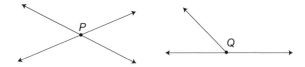

isosceles triangle: A triangle with at least two sides the same length.

justify: To prove or show to be true or valid using logic and/or evidence.

key: An explanation of what each symbol represents in a pictograph.

Glossary

kilometer (km): A metric unit of length: 1 kilometer = 1,000 meters.

line: A straight path of points that goes on forever in both directions.

line graph: A graph that uses a line or a curve to show how data changes over time.

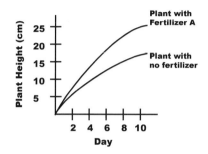

line of symmetry: A line on which a figure can be folded into two parts so that the parts match exactly.

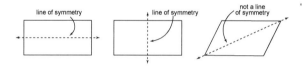

liter (L): A metric unit of capacity: 1 liter = 1,000 milliliters.

mass: The amount of matter an object has.

mean: Also called arithmetic average. A number found by adding two or more quantities together, and then dividing the sum by the number of quantities. For example, in the set {9, 5, 4} the mean is 6: 9 + 5 + 4 = 18; 18 ÷ 3 = 6. *See average.*

median: The middle number when numbers are put in order from least to greatest or from greatest to least. For example, in the set of numbers 6, 7, 8, 9, 10, the number 8 is the median (middle) number.

meter (m): A metric unit of length: 1 meter = 100 centimeters.

method: A systematic way of accomplishing a task.

mixed number: A number consisting of a whole number and a fraction. **Example:** $6\frac{2}{3}$.

mode: The number or numbers that occur most often in a set of data. Example: The mode of {1, 3, 4, 5, 5, 7, 9} is 5.

multiple: A product of a number and any other whole number. Examples: {2, 4, 6, 8, 10, 12,…} are multiples of 2.

multiplication: An operation on two numbers that tells how many in all. The first number is the number of sets and the second number tells how many in each set.

number line: A line that shows numbers in order using a scale. Equal intervals are marked and usually labeled on the number line.

number sentence: An expression of a relationship between quantities as an equation or an inequality. Examples: 7 + 7 = 8 + 6; 14 < 92; 56 + 4 > 59.

numerator: The top number in a fraction.

octagon: An eight-sided polygon. The total measure of the angles within an octagon is 1080°.

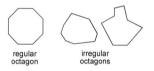

odd number: A whole number that has 1, 3, 5, 7, or 9 in the ones place. An odd number is not divisible by 2. Examples: The numbers 53 and 701 are odd numbers.

operation: A mathematical process that combines numbers; basic operations of arithmetic include addition, subtraction, multiplication, and division.

order: To arrange numbers from the least to greatest or from the greatest to least.

ordered pair: Two numbers inside a set of parentheses separated by a comma that are used to name a point on a coordinate grid.

parallel lines: Lines in the same plane that never intersect.

parallelogram: A quadrilateral in which opposite sides are parallel.

pattern: An arrangement of numbers, pictures, etc., in an organized and predictable way. Examples: 3, 6, 9, 12, or ®0®0®0.

Glossary

pentagon: A five-sided polygon. The total measure of the angles within a pentagon is 540°.

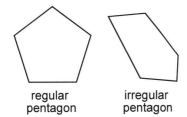

regular pentagon   irregular pentagon

perimeter: The distance around a figure.

perpendicular lines: Two lines that intersect to form a right angle (90 degrees).

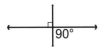

pictograph: A graph that uses pictures or symbols to represent similar data. The value of each picture is interpreted by a "key" or "legend."

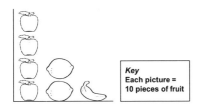

place value: The value given to the place a digit has in a number.
Example: In the number 135, the 1 is in the hundreds place so it represents 100 (1 x 100); the 3 is in the tens place so it represents 30 (3 x 10); and the 5 is in the ones place so it represents 5 (5 x 1).

p.m.: The hours from noon to midnight; from the Latin words *post meridiem* meaning "after noon."

point: An exact position often marked by a dot.

polygon: A closed figure made up of straight line segments.

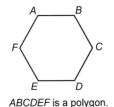

ABCDEF is a polygon.

possible event: An event that might or might not happen.

predict: To tell what you believe may happen in the future.

prediction: A prediction is a description of what may happen before it happens.

probability: The likelihood that something will happen.

product: The answer to a multiplication problem.
Example: In 3 x 4 = 12, 12 is the product.

pyramid: A solid figure in which the base is a polygon and faces are triangles with a common point called a vertex.

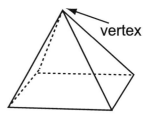

quadrilateral: A four-sided polygon. Rectangles, squares, parallelograms, rhombi, and trapezoids are all quadrilaterals. The total measure of the angles within a quadrilateral is 360°.
Example: ABCD is a quadrilateral.

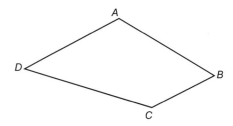

questionnaire: A set of questions for a survey.

quotient: The answer in a division problem.
Dividend ÷ divisor = quotient.
Example: In 15 ÷ 3 = 5, 5 is the quotient.

range: The difference between the least number and the greatest number in a data set. For example, in the set {4, 7, 10, 12, 36, 7, 2}, the range is 34; the greatest number (36) minus the least number (2): (36 – 2 = 34).

Glossary

rectangle: A quadrilateral with four right angles. A square is one example of a rectangle.

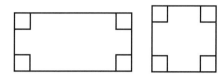

reflection: The change in the position of a figure that is the result of picking it up and turning it over. *See flip.*

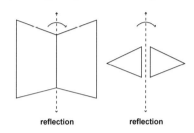

remainder: The number that is left over after dividing. Example: In 31 ÷ 7 = 4 R 3, the 3 is the remainder.

represent: To present clearly; describe; show.

rhombus: A quadrilateral with opposite sides parallel and all sides the same length. A square is one kind of rhombus.

right angle: An angle that forms a square corner and measures 90 degrees.

right triangle: A triangle having one right angle. *See right angle and triangle.*

rounding: Replacing an exact number with a number that tells about how much or how many to the nearest ten, hundred, thousand, and so on.
Example: 52 rounded to the nearest 10 is 50.

rule: A procedure; a prescribed method; a way of describing the relationship between two sets of numbers. Example: In the following data, the rule is to add 3:

Input	Output
3	6
5	8
9	12

ruler: A straight-edged instrument used for measuring the lengths of objects. A ruler usually measures smaller units of length, such as inches or centimeters.

scale: The numbers that show the size of the units used on a graph.

sequence: A set of numbers arranged in a special order or pattern.

set: A group made up of numbers, figures, or parts.

side: A line segment connected to other segments to form the boundary of a polygon.

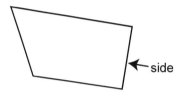

similar: A description for figures that have the same shape.

slide (translation): The change in the position of a figure that moves up, down, or sideways. Example: scooting a book on a table.

solids: Figures in three dimensions.

solve: To find the solution to an equation or problem; finding the values of unknown variables that will make a true mathematical statement.

sphere: A solid figure in the shape of a ball. Example: a basketball is a sphere.

square: A rectangle with congruent (equal) sides. *See rectangle.*

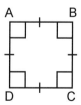

square number: The product of a number multiplied by itself.
Example: 49 is a square number (7 x 7 = 49).

square unit: A square with sides 1 unit long, used to measure area.

Glossary

standard form: A way to write a number showing only its digits. Example: 2,389.

standard units of measure: Units of measure commonly used; generally classified in the U.S. as the customary system or the metric system:

> **Customary System:**
> **Length**
> 1 foot (ft) = 12 inches (in)
> 1 yard (yd) = 3 feet or 36 inches
> 1 mile (mi) = 1,760 yards or 5,280 feet
>
> **Weight**
> 16 ounces (oz) = 1 pound (lb)
> 2,000 pounds = 1 ton (t)
>
> **Capacity**
> 1 pint (pt) = 2 cups (c)
> 1 quart (qt) = 2 pints
> 1 gallon (gal) = 4 quarts

> **Metric System:**
> **Length**
> 1 centimeter (cm) = 10 millimeters (mm)
> 1 decimeter (dm) = 10 centimeters
> 1 meter (m) = 100 centimeters
> 1 kilometer (km) = 1,000 meters
>
> **Weight**
> 1,000 milligrams (mg) = 1 gram (g)
> 1,000 grams (g) = 1 kilogram (kg)
> 1,000 kilograms (kg) = 1 tonne (metric ton)
>
> **Capacity**
> 1 liter (l) = 1,000 milliliters (ml)

strategy: A plan used in problem solving, such as looking for a pattern, drawing a diagram, working backward, etc.

subtraction: The operation that finds the difference between two numbers.

sum: The answer when adding two or more addends. Addend + Addend = Sum.

summary: A series of statements containing evidence, facts, and/or procedures that support a result.

survey: A way to collect data by asking a certain number of people the same question and recording their answers.

symmetry: A figure has line symmetry if it can be folded along a line so that both parts match exactly. A figure has radial or rotational symmetry if, after a rotation of less than 360°, it is indistinguishable from its former image.

Examples of Figures With At Least Two Lines of Symmetry

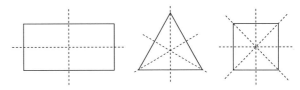

table: A method of displaying data in rows and columns.

temperature: A measure of hot or cold in degrees.

translation (slide): A change in the position of a figure that moves it up, down, or sideways.

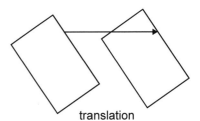

translation

triangle: A polygon with three sides. The sum of the angles of a triangle is always equal to 180°.

turn: The change in the position of a figure that moves it around a point. Also called a rotation. Example: The hands of a clock turn around the center of the clock in a clockwise direction.

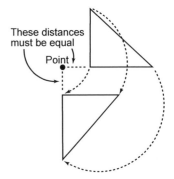

These distances must be equal
Point

Glossary

unlikely event: An event that probably will not happen.

vertex: The point where two rays meet to form an angle or where the sides of a polygon meet, or the point where 3 or more edges meet in a solid figure.

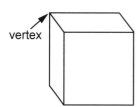

whole number: An integer in the set {0, 1, 2, 3 . . .}. In other words, a whole number is any number used when counting in addition to zero.

word forms: The number written in words. Examples: 546 is "five hundred forty-six."

Examples of Common Two-Dimensional Geometric Shapes

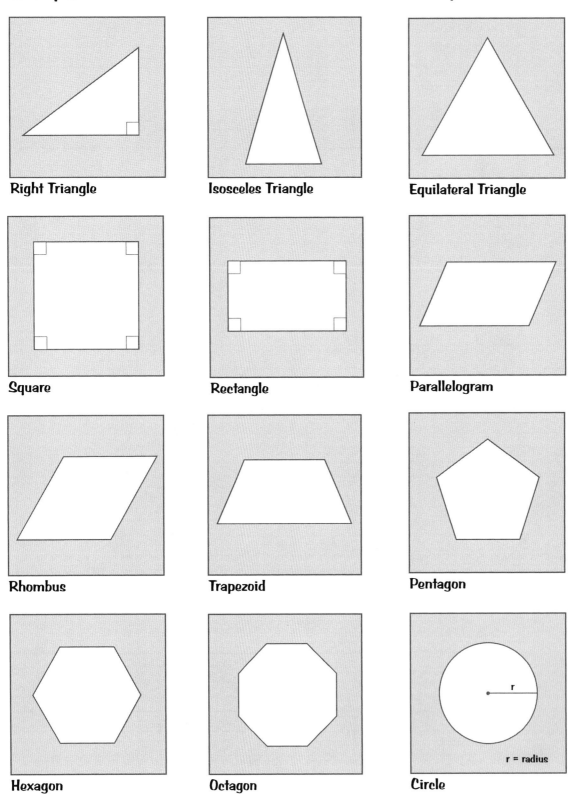

Examples of How Lines Interact

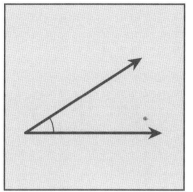
Acute Angle

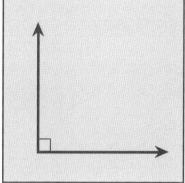
Right Angle

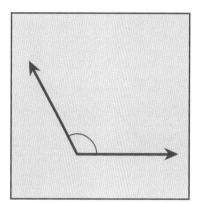
Obtuse Angle

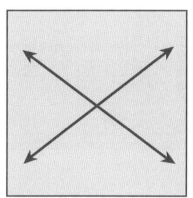
Intersecting

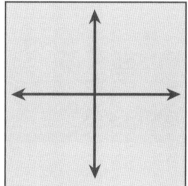
Perpendicular

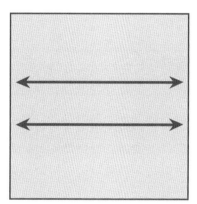
Parallel

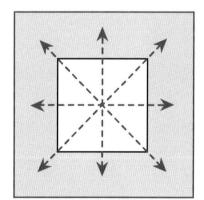
Lines of Symmetry

Examples of Common Types of Graphs

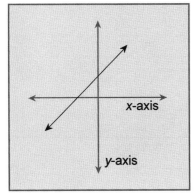
Line Graph

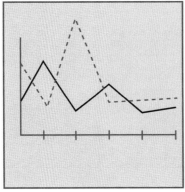
Double Line Graph

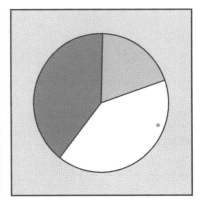
Pie Chart

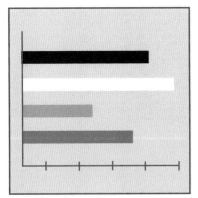
Bar Graph

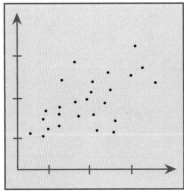
Scatterplot

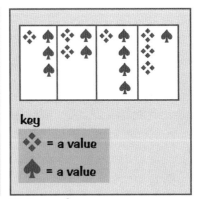
Pictograph

Examples of Common Three-Dimensional Objects

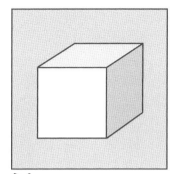
Cube

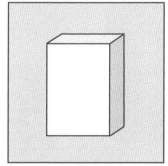
Rectangular Prism

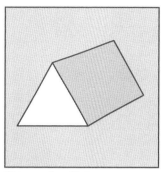
Triangular Prism

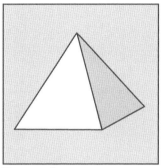
Pyramid

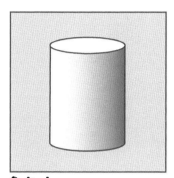
Cylinder

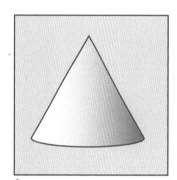
Cone

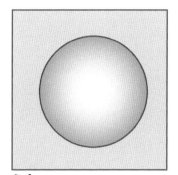
Sphere

Examples of Object Movement

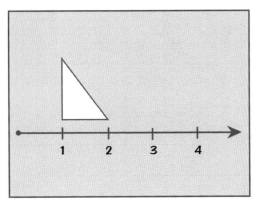 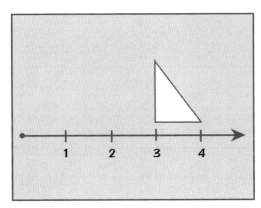

Translation

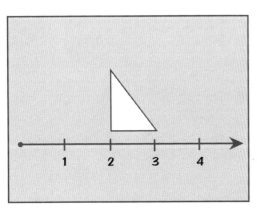 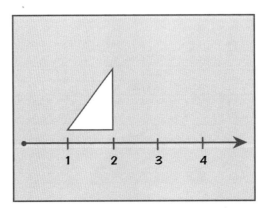

Reflection

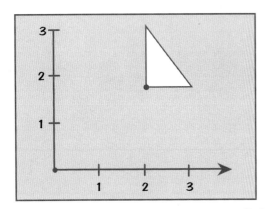 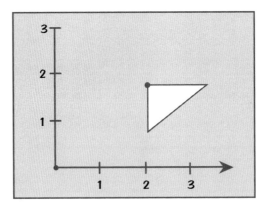

Rotation

Mathematics Assessment One

Directions for Taking the Mathematics Assessment

The Mathematics Assessment is made up of multiple-choice and short-answer questions. These questions show you how the skills you have learned in Mathematics class may be tested. The questions also give you a chance to practice your skills. If you have trouble with an area, talk with a parent or teacher.

Multiple-choice questions require you to pick the best answer out of three possible choices. Only one answer is correct. The short-answer questions will ask you to write your answer and explain your thinking using words, numbers, or pictures, or to show the steps you used to solve a problem. Remember to read the questions and the answer choices carefully. You will mark your answers on the answer document.

When you finish, check your answers.

1. Look at the equation.

$$4 \times 5 = \boxed{}$$

Which story matches the equation?

A. In the candy shop there are 10 types of fancy chocolates. There are 4 rows of chocolate creams. How many more rows of chocolate creams are needed to have 5 rows altogether?

B. In the candy shop there are 10 types of fancy chocolates. There are 4 rows of chocolate creams with 5 candies in each row. How many chocolate creams are there altogether?

C. In the candy shop there are 10 types of fancy chocolates. There are 4 rows of chocolate creams. There are 5 rows of caramels. How many rows of fancy chocolates are there altogether?

2. Rob had 12 seeds and 3 pots to put them into. He put the same number of seeds into each pot.

Which number sentence shows how to find the number of seeds he put into each pot?

A. $12 \div 3$

B. 12×3

C. $12 + 3$

3. Division is the same as taking away equal groups.

   The picture shows the number sentence 12 ÷ 3 = 4. When you have 12 stars, you can take away 4 equal groups of 3 stars.

   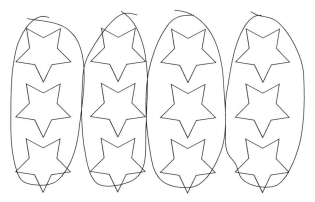

   Draw a picture to show 15 ÷ 5 = 3.

4. Cheryl had 30 dimes. She divided them into piles with 5 dimes in each pile.

   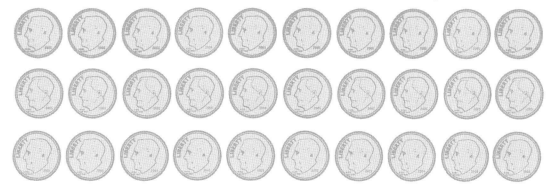

   How many piles were there?

   A. 5 piles

   B. 6 piles

   C. 10 piles

5. Devon has 7 flowerpots. She wants to plant 4 flower seeds in each pot.

How many seeds will she need?

A. 21 seeds

B. 24 seeds

C. 28 seeds

6. Which of the expressions below correctly describes the number represented by □ in the equation 18 ÷ □ = 6?

A. □ is equal to 18

B. □ is less than 18

C. □ is greater than 18

7. In some math operations order is not important.

Since 3 × 5 = 15 and 5 × 3 = 15, then 3 × 5 = 5 × 3.

Which of the following statements is also true?

A. 8 ÷ 2 = 2 ÷ 8

B. 8 − 5 = 5 − 8

C. 7 + 2 = 2 + 7

8. Which of the following number sentences can be completed using the number 8 so it is in the same fact family as 4 × 8 = 32?

A. 32 ÷ □ = 4

B. 3 × □ = 24

C. 24 + □ = 32

9. Which group of number sentences is all in the same fact family?

   A. 5 × 6 = 30, 30 ÷ 6 = 5, 6 × 5 = 30, 30 ÷ 5 = 6

   B. 5 × 6 = 30, 5 × 7 = 35, 5 × 8 = 40, 5 × 9 = 45

   C. 5 × 6 = 30, 30 ÷ 6 = 5, 10 × 3 = 30, 30 ÷ 3 = 10

10. There are 645 people at the school fair. Of those 645 people, 417 of them are students. There are also 67 teachers at the fair. The rest of the people at the fair are parents.

    How many parents are at the school fair?

    A. 128 parents

    B. 161 parents

    C. 228 parents

11. The golf team is going to an away match. There are 10 players on the team. The players will ride in cars. Each car can hold 3 players and 1 driver.

    What is the **fewest** number of cars they will need to take everyone to the match?

    Show your work using words, numbers, or pictures.

12. Mr. Ying has 7 pots of tulips in his shop. Each pot has 7 flowers in it. The chart below shows how many tulips he has in 5 pots.

Pots	Tulips
1	7
2	14
3	21
4	28
5	35
6	
7	
8	

    How many tulips does he have in all 7 pots?

    A. 53 tulips

    B. 51 tulips

    C. 49 tulips

13. Federico counted 35 apples on an apple tree.

    About how many apples are on the apple tree?

    A. 50 apples

    B. 40 apples

    C. 30 apples

14. Dan ran 43 miles last week and 51 miles this week.

    How many **more** miles did Dan run this week?

    A. 8 miles

    B. 11 miles

    C. 12 miles

15. What is 3 × 60?

    A. 18

    B. 90

    C. 180

16. What fraction of the rectangle below is shaded?

    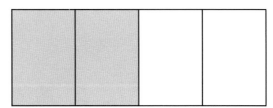

    A. $\frac{1}{4}$

    B. $\frac{1}{2}$

    C. $\frac{4}{5}$

17. Addie colored $\frac{2}{3}$ of her picture.

    Which picture shows $\frac{2}{3}$ colored?

    A.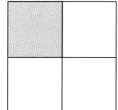

    B.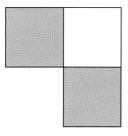

    C.

18. Which fraction is shown on the number line?

    A. $\frac{1}{4}$

    B. $\frac{2}{3}$

    C. $\frac{3}{4}$

19. Sean ate $\frac{1}{2}$ of the cheese pizza and Trevor ate $\frac{1}{4}$ of the meatball pizza. Which of the following statements is true? Mark your answer.

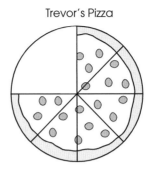

A. Sean ate less pizza than Trevor ate.

B. Sean ate more pizza than Trevor ate.

C. Trevor ate $\frac{1}{4}$ more pizza than Sean ate.

20. Ursula opened her sock drawer and saw she had a total of 15 socks in her drawer. Of the 15 socks she had, 5 of them were blue, meaning $\frac{5}{15}$ of her socks were blue.

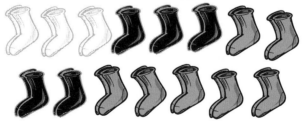

Which of the following fractions is equal to $\frac{5}{15}$?

A. $\frac{1}{5}$

B. $\frac{1}{3}$

C. $\frac{3}{5}$

21. Carrie set her alarm clock for 7:30 a.m. Which of the clocks below shows the time 7:30?

   A.

   B.

   C.

22. What time does the clock below show?

   A. 8:15

   B. 8:45

   C. 9:20

23. On the balance scale below, two striped marbles are needed to equal one solid silver marble.

If one striped marble weighs 4 grams, how many ounces does a silver marble weigh?

- A. 2 grams
- B. 6 grams
- C. 8 grams

24. Four students in Mrs. Campbell's class won blue ribbons on Field Day. Jessica won 2 ribbons, Phillip and Marcus won 4 ribbons each, and Sun won 3 ribbons.

Which pictograph shows this information?

A. Field Day Awards

Jessica	🎗
Sun	🎗 🎗
Phillip	🎗 🎗 (half)
Marcus	🎗 🎗

Each 🎗 represents 2 ribbons.

B. Field Day Awards

Jessica	🎗 🎗
Sun	🎗 🎗 🎗 🎗
Phillip	🎗 🎗 🎗
Marcus	🎗 🎗 🎗 🎗

Each 🎗 represents 2 ribbons.

C. Field Day Awards

Jessica	🎗
Sun	🎗 🎗 (half)
Phillip	🎗 🎗
Marcus	🎗 🎗

Each 🎗 represents 2 ribbons.

25. Audrey is selling candy bars to help raise money for her school. The bar graph below shows how many candy bars she sold in one week.

Which of the following statements is true?

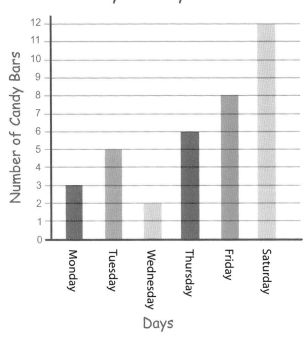

A. Audrey sold twice as many candy bars on Saturday as she did on Thursday.

B. Audrey sold three more candy bars on Tuesday than she did on Monday.

C. Audrey sold three times as many candy bars on Friday as she did on Wednesday.

26. The students in Mr. Sefton's class took a survey to see what holiday was most popular with the students in their class. They put their results into the bar graph below.

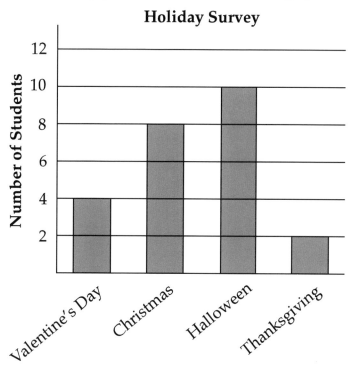

Use the information in the bar graph to find the total number of students in Mr. Sefton's class. Write a number sentence to tell how many students are in Mr. Sefton's class.

27. If you wanted to find the area of your desktop, what would be an appropriate unit of measurement to use?

   A. Feet
   B. Square feet
   C. Square inches

28. What is the area of the flag below in square units?

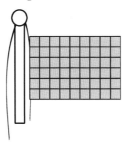

A. 58 square units

B. 54 square units

C. 48 square units

29. Phil is building a new patio. The patio will be 5 meters long and 4 meters wide.

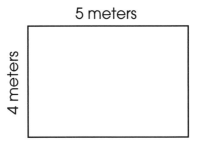

What will the area of the new patio be?

A. 9 square meters

B. 18 square meters

C. 20 square meters

30. Max plans on putting a fence around his garden.

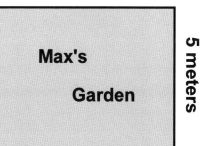

What is the perimeter of the garden?

A. 16 meters

B. 22 meters

C. 23 meters

31. Sam walked all the way around his block. His block is a rectangle that is 100 yards long and 50 yards wide.

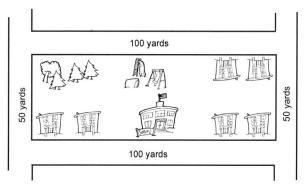

How far did Sam walk?

On your own paper, show how you got your answer using words, numbers, or pictures.

Go On

32. Tyler is playing a game. He must find the quadrilateral that fits the clues.

> **Clues**
> The mystery quadrilateral is a closed shape with four straight sides.
> The mystery quadrilateral has sides that are all the same length.
> The mystery quadrilateral has four right angles.

What is the mystery quadrilateral?

A.

B.

C.

33. The rectangle below has an area of 16 square units.

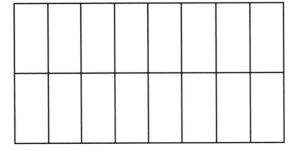

Draw lines on the rectangle to divide it into 4 equal parts.

What is the area of one of the equal parts?

34. For three weeks, the students at Kenwood School sold tickets to their spring play. The first week they sold 124 tickets, the second week they sold 96 tickets, and the third week they sold 153 tickets.

    Rounding to the nearest hundred, about how many tickets did the students sell altogether?

    A. 200

    B. 300

    C. 400

35. Ben and his mother baked 36 cookies. They ate 5 cookies and gave 12 cookies to a neighbor.

    How many cookies do they have left?

    A. 17

    B. 19

    C. 24

36. Sarah is looking in magazines for multiples of three to use on her math poster.

    Which numbers will Sarah use?

    A. 3, 6, and 9

    B. 12, 16, and 18

    C. 15, 16, and 18

37. Which square has an area of 16 square inches?

   A. 4 inches × 4 inches square

   B. 4 cm × 4 cm square

   C. 8 inches × 8 inches square

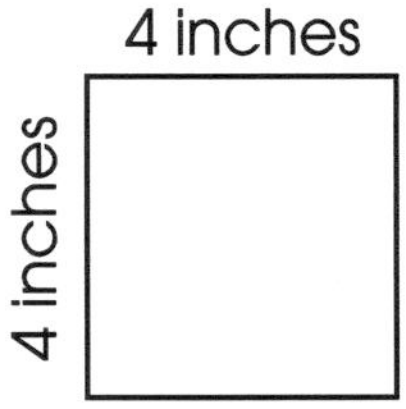

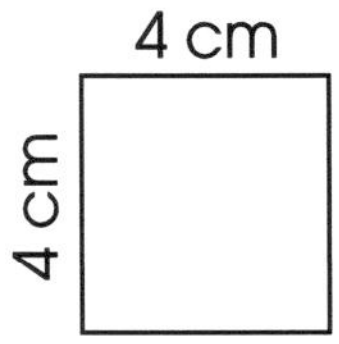

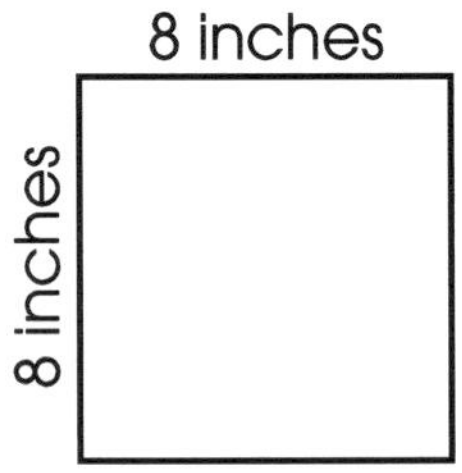

38. What fraction of the flowers below have petals that are shaded?

A. $\frac{4}{6}$

B. $\frac{4}{8}$

C. $\frac{6}{3}$

39. Look at the equation.

$$15 \div \boxed{} = 5$$

Which story matches the equation?

A. Tom buys 5 pieces of candy. Then he buys 10 more pieces. How many pieces of candy does he have now?

B. Tom buys 15 pieces of candy. Then he shares the candy with his friends. Tom gives each friend 5 pieces of candy. How many friends did Tom share with?

C. Tom eats 10 pieces of candy. He has 5 pieces left. How many pieces of candy did he have to start with?

40. Kevin has 32 model cars and 17 model trucks.

   About how many cars and trucks does he have altogether? Estimate to the nearest 10.

   A. 40
   B. 50
   C. 60

Mathematics Assessment One
Answer Document

1. Ⓐ Ⓑ Ⓒ
2. Ⓐ Ⓑ Ⓒ
3.

4. Ⓐ Ⓑ Ⓒ
5. Ⓐ Ⓑ Ⓒ
6. Ⓐ Ⓑ Ⓒ
7. Ⓐ Ⓑ Ⓒ
8. Ⓐ Ⓑ Ⓒ
9. Ⓐ Ⓑ Ⓒ
10. Ⓐ Ⓑ Ⓒ

Mathematics Assessment One
Answer Document

11 [blank response box]

12 Ⓐ Ⓑ Ⓒ
13 Ⓐ Ⓑ Ⓒ
14 Ⓐ Ⓑ Ⓒ
15 Ⓐ Ⓑ Ⓒ
16 Ⓐ Ⓑ Ⓒ
17 Ⓐ Ⓑ Ⓒ
18 Ⓐ Ⓑ Ⓒ
19 Ⓐ Ⓑ Ⓒ
20 Ⓐ Ⓑ Ⓒ
21 Ⓐ Ⓑ Ⓒ
22 Ⓐ Ⓑ Ⓒ
23 Ⓐ Ⓑ Ⓒ
24 Ⓐ Ⓑ Ⓒ
25 Ⓐ Ⓑ Ⓒ

26

27 Ⓐ Ⓑ Ⓒ

28 Ⓐ Ⓑ Ⓒ

29 Ⓐ Ⓑ Ⓒ

30 Ⓐ Ⓑ Ⓒ

31

Mathematics Assessment One
Answer Document

32 Ⓐ Ⓑ Ⓒ

33

The area of one of the equal parts is _____ .

34 Ⓐ Ⓑ Ⓒ
35 Ⓐ Ⓑ Ⓒ
36 Ⓐ Ⓑ Ⓒ
37 Ⓐ Ⓑ Ⓒ
38 Ⓐ Ⓑ Ⓒ
39 Ⓐ Ⓑ Ⓒ
40 Ⓐ Ⓑ Ⓒ

Mathematics Assessment Two

Directions for Taking the Mathematics Assessment

The Mathematics Assessment is made up of multiple-choice and short-answer questions. These questions show you how the skills you have learned in mathematics class may be tested. The questions also give you a chance to practice your skills. If you have trouble with an area, talk with a parent or teacher.

Multiple-choice questions require you to pick the best answer out of three possible choices. Only one answer is correct. The short-answer questions will ask you to write your answer and explain your thinking using words, numbers, or pictures, or to show the steps you used to solve a problem. Remember to read the questions and the answer choices carefully. You will mark your answers on the answer document.

When you finish, check your answers.

1. Mr. St. Meyers arranged the playground balls into piles in a pattern as shown below.

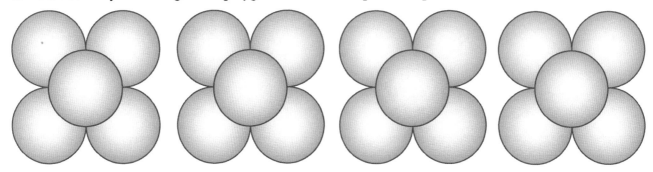

Which number sentence best represents the pattern Mr. St. Meyers used?

A. 4 x 3 = 12

B. 4 x 5 = 20

C. 4 + 5 = 9

2. Look at the number sentence.

$$24 \div 3 = 8$$

Write one sentence to describe a real-world problem that could be solved using this number sentence.

3. On a recent fishing trip, the Summers family caught 25 fish.

   If each of the 5 members of the family caught an equal number of fish, which of the following could be used to find the number of fish each member of the family caught?

   A. 25 + 5

   B. 25 − 5

   C. 25 ÷ 5

4. Mrs. Garcia has 3 cupcake pans. Each pan has room for 9 cupcakes.

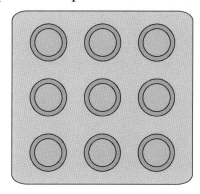

   How many cupcakes can Mrs. Garcia bake at one time?

   A. 15 cupcakes

   B. 18 cupcakes

   C. 27 cupcakes

5. Cheryl had 30 star stickers. She divided them into 5 equal piles.

How many star stickers were in each pile?

A.

B.

C.

6. Look at the number sentence.

$$4 \times \boxed{} = 0$$

Which of the numbers below will make this number sentence true?

A. 0

B. 1

C. 4

7. Look at the equation.

    2 x 3 x 2 = 12

    Which of the following equations is equal to the equation above?

    A. 3 x 2 x 2 = 12
    B. 3 x 2 x 3 = 12
    C. 3 x 2 x 12 = 2

8. Which of the following number sentences is in the same fact family as 6 x 9 = 54?

    A. ☐ x 6 = 36
    B. ☐ x 9 = 36
    C. ☐ ÷ 9 = 6

9. Look at the number sentence.

    5 x 6 = 30

    Which number sentence is in the same fact family as 5 x 6 = 30?

    A. 30 ÷ 6 = 5
    B. 36 ÷ 6 = 6
    C. 5 x 5 = 25

10. Shauna has 78 rocks in her rock collection. During the summer, she collects 17 more rocks. Then she gives 12 rocks to one of her friends.

    About how many rocks does Shauna have now?

    A. 90
    B. 100
    C. 110

11. Lindsey has $5.00. She wants to order a sandwich, drink, and ice cream from the lunch menu:

Does Lindsey have enough money for this lunch?

Show or explain your work using words, numbers, and/or pictures.

12. The chart below shows how many apples are needed to make jars of applesauce.

Jars	1	2	3	4
Apples	8	16	24	32

According to the pattern in the chart, how many apples would be used to make 6 jars of applesauce?

A. 32 apples

B. 40 apples

C. 48 apples

13. The wallpaper in Lisa's room has lions on it. On one wall, Lisa counted 39 lions.

    About how many lions are on that wall?

    A. 30

    B. 35

    C. 40

14. The Atlantic Coast of Florida has 399 miles of shoreline and the Gulf Coast has 798 miles of shoreline.

    How many **more** miles of shoreline does the Gulf Coast have than the Atlantic Coast?

    A. 399 miles

    B. 401 miles

    C. 798 miles

15. What is 4 × 50?

    A. 20

    B. 90

    C. 200

16. Which figure has $\frac{5}{6}$ shaded?

    A.
    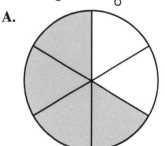

    B.

    C.
    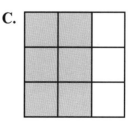

17. Look at the stars below.

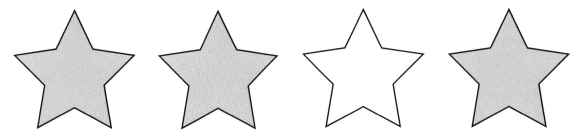

Which of the fractions below shows the stars that are shaded?

A. $\frac{1}{4}$

B. $\frac{2}{3}$

C. $\frac{3}{4}$

18. Which fraction is shown on the number line?

A. $\frac{1}{5}$

B. $\frac{2}{6}$

C. $\frac{1}{6}$

19. Carlos and Tamir ordered 2 large pizzas. Carlos ate $\frac{1}{8}$ of his pizza and Tamir ate $\frac{1}{4}$ of his pizza.

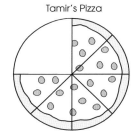

Who ate more pizza?

A. Carlos ate more pizza than Tamir ate.

B. Tamir ate more pizza than Carlos ate.

C. Carlos and Tamir ate the same amount of pizza.

20. Look at the model.

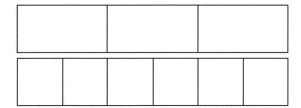

Using the model, which fraction is equal to $\frac{1}{3}$?

A. $\frac{1}{6}$

B. $\frac{2}{6}$

C. $\frac{3}{6}$

21. Which clock matches the time shown on the digital clock below?

A.

B.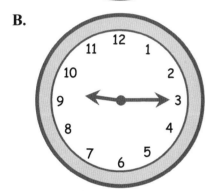

C.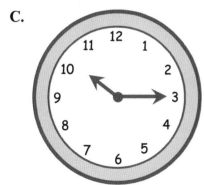

22. What time is shown on the clock below?

A. 3:40

B. 3:45

C. 4:45

23. Which of the following is the approximate weight of a 20-pound tire?

Key
1 gram = .002 pounds
1 kilogram = 2.2 pounds

A. 90 grams

B. 9 kilograms

C. 90 kilograms

24. Tara has 12 pop CDs, 6 classical CDs, 4 jazz CDs, and 10 CDs with show tunes. Which of the graphs below shows Tara's CD collection?

A.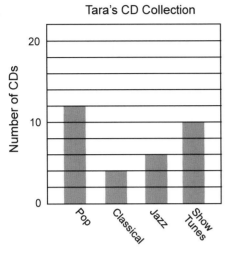

B.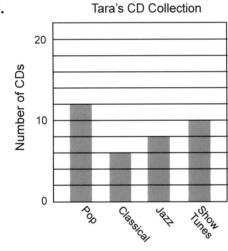

C.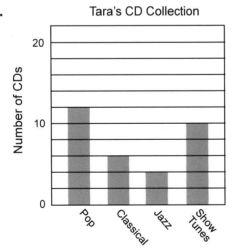

25. Rita has plants in 3 rooms of her house. The graph below shows how many plants are in each room.

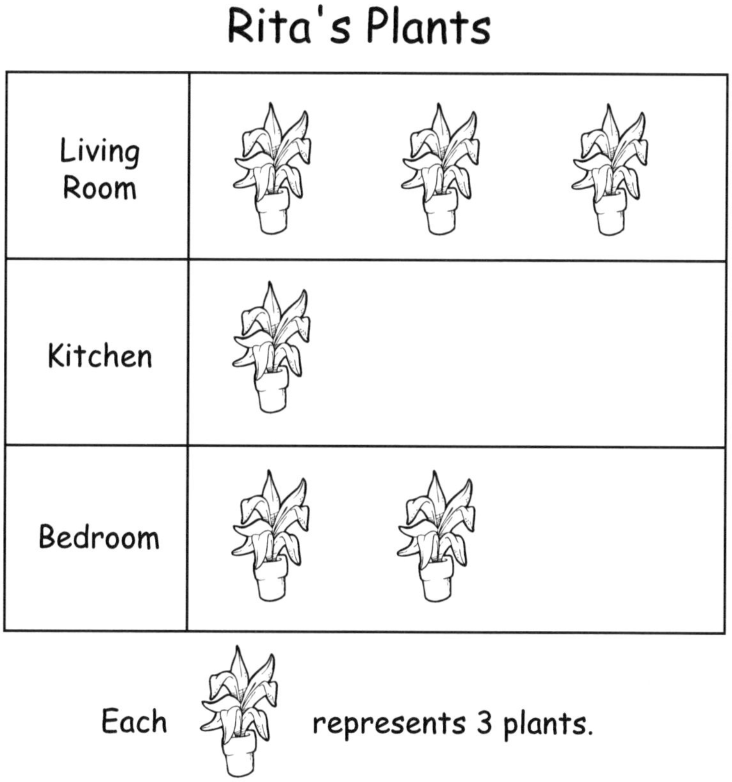

Rita's Plants

Living Room	🌱 🌱 🌱
Kitchen	🌱
Bedroom	🌱 🌱

Each 🌱 represents 3 plants.

How many plants does Rita have in her house?

A. 6 plants

B. 12 plants

C. 18 plants

26. Carly takes a survey of her class to find out what pets they have. She makes a tally mark for every pet.

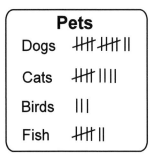

Draw a bar graph to show the results of Carly's survey.

- Write a title for the bar graph.
- Label and number the scale.
- Draw the bars.
- Label each bar.
- Include a label for all the bars together.

27. **About** how long is this caterpillar?

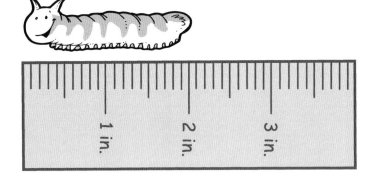

A. 1 in.

B. $1\frac{1}{2}$ in.

C. 2 in.

28. The grid below shows the size of Jasmine's room. Her bed is shown in the room.

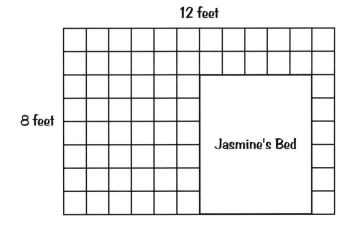

What is the area of Jasmine's bed?

A. 20 square feet

B. 22 square feet

C. 30 square feet

29. Ana is hanging posters in her room. Her poster is 5 feet long by 3 feet wide.

What is the area of the wall that her poster will cover?

A. 8 square feet

B. 15 square feet

C. 16 square feet

30. Mr. Kahn has a rectangular swimming pool, as shown below.

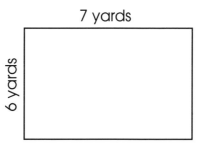

How would Mr. Kahn find the perimeter of his pool?

A. Add 7 yards, 7 yards, 6 yards, and 6 yards.

B. Add 7 yards and 6 yards.

C. Multiply 7 yards and 6 yards.

31. Tennis is played on a rectangular court like the one shown below.

What is the perimeter of the tennis court?

Show your work using words, numbers, or pictures.

32. Which drawing shows a closed shape with four straight sides, all the same length, and four right angles?

A.

B.

C.

33. The rectangle below has an area of 24 square units.

Draw lines on the rectangle to divide it into 3 equal parts.

What is the area of one of the equal parts?

34. Chen wants to use estimation to check her answer to this addition problem:

$$126 + 245 + 283 + 379 = \square$$

She checks her answer by rounding each number to the nearest hundred and then adding.

Which of these choices shows how Chen estimated the sum?

A. $100 + 200 + 200 + 300 = 800$

B. $100 + 200 + 300 + 400 = 1{,}000$

C. $100 + 200 + 300 + 300 = 900$

35. Kris has 4 rosebushes in her yard. She picks 3 roses from each bush. She wants to put the roses into 2 vases, with the same number of roses in each vase.

How many roses will she put into each vase?

A. 3

B. 6

C. 12

36. There are 3 wheels on a tricycle.

Which of the charts below shows this information?

A.

Tricycles	Wheels
3	3
6	6
9	9

B.

Tricycles	Wheels
1	3
2	6
3	9

C.

Tricycles	Wheels
1	3
3	6
6	12

Go On

37. In each of the figures shown below, the line segments are all the same length.

    Which figure has a perimeter of 18 cm?

    A.
    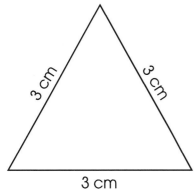

    B.
    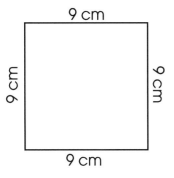

    C.
    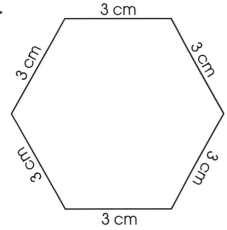

38. A wall of a house is built with equal-sized bricks. Each brick is about 2 inches wide and 6 inches long.

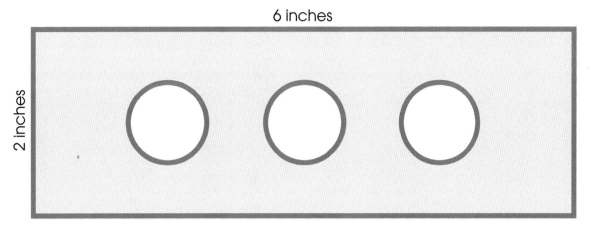

If 8 bricks are placed in a row so that they are lying down like the brick above, what will the total length be?

A. 36 inches

B. 40 inches

C. 48 inches

39. Zack sketched 3 pictures a day for 5 days on his drawing pad.

    Which number sentence shows the correct way to find the total number of pictures Zack sketched over 5 days?

    A. 3 + 5

    B. 3 × 5

    C. 5 ÷ 3

40. Mrs. Chen's class took a field trip to the zoo. She has 27 students in her class. When they arrived at the zoo, they divided into two groups. Group A had 12 students in it.

    How many students were in Group B?

    A. 13 students

    B. 14 students

    C. 15 students

1. Ⓐ Ⓑ Ⓒ

2.

3. Ⓐ Ⓑ Ⓒ
4. Ⓐ Ⓑ Ⓒ
5. Ⓐ Ⓑ Ⓒ
6. Ⓐ Ⓑ Ⓒ
7. Ⓐ Ⓑ Ⓒ
8. Ⓐ Ⓑ Ⓒ
9. Ⓐ Ⓑ Ⓒ
10. Ⓐ Ⓑ Ⓒ

Mathematics Assessment Two
Answer Document

11.

[blank response box]

12. Ⓐ Ⓑ Ⓒ
13. Ⓐ Ⓑ Ⓒ
14. Ⓐ Ⓑ Ⓒ
15. Ⓐ Ⓑ Ⓒ
16. Ⓐ Ⓑ Ⓒ
17. Ⓐ Ⓑ Ⓒ
18. Ⓐ Ⓑ Ⓒ
19. Ⓐ Ⓑ Ⓒ
20. Ⓐ Ⓑ Ⓒ
21. Ⓐ Ⓑ Ⓒ
22. Ⓐ Ⓑ Ⓒ
23. Ⓐ Ⓑ Ⓒ
24. Ⓐ Ⓑ Ⓒ
25. Ⓐ Ⓑ Ⓒ

26

27 Ⓐ Ⓑ Ⓒ
28 Ⓐ Ⓑ Ⓒ
29 Ⓐ Ⓑ Ⓒ
30 Ⓐ Ⓑ Ⓒ

31

Mathematics Assessment Two
Answer Document

Show What You Know® on the Common Core for Grade 3

32 Ⓐ Ⓑ Ⓒ

33

The area of one of the equal parts is _____ .

34 Ⓐ Ⓑ Ⓒ
35 Ⓐ Ⓑ Ⓒ
36 Ⓐ Ⓑ Ⓒ
37 Ⓐ Ⓑ Ⓒ
38 Ⓐ Ⓑ Ⓒ
39 Ⓐ Ⓑ Ⓒ
40 Ⓐ Ⓑ Ⓒ

Notes

Notes

Notes

Notes

Notes

Show What You Know® on the COMMON CORE

Assessing Student Knowledge of the Common Core State Standards (CCSS)
Reading • Mathematics • Grades 3–8

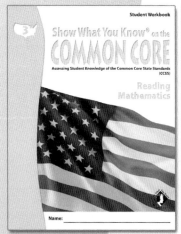 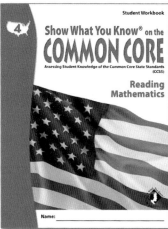 

Diagnostic Test-Preparation Student Workbooks and Parent/Teacher Editions for Grades 3–5

 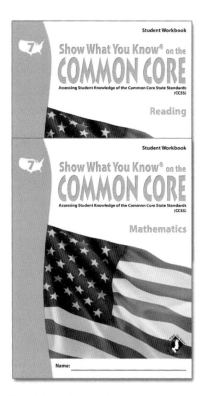 

Single Subject Student Workbooks and Parent/Teacher Editions for Grades 6–8

**For More Information, call our toll-free number: 1.877.PASSING (727.7464)
or visit our website: www.showwhatyouknowpublishing.com**